ONSHORE IMPACTS OF OFFSHORE OIL

Proceedings of the International Conference on Oil and the Environment, Scotland 1980, held in the Pollock Halls, University of Edinburgh, 28 September – 1 October. Conference chairmen: Alastair M. Dunnett (Director, Thomson Scottish Petroleum, Edinburgh, UK); Egil Ellingsen (President, Norske Petroleumsforening (Norwegian Petroleum Society), Oslo, Norway). Hon. conference director: William J. Cairns. Hon. conference managers: W. J. Cairns and Associates (Overseas) Ltd, Edinburgh, UK.

ONSHORE IMPACTS OF OFFSHORE OIL

Edited by

WILLIAM J. CAIRNS and PATRICK M. ROGERS

W. J. Cairns and Partners, Environmental Consultants
16 Randolph Crescent, Edinburgh EH3 7TT, UK

APPLIED SCIENCE PUBLISHERS LTD
LONDON and NEW JERSEY

APPLIED SCIENCE PUBLISHERS LTD
Ripple Road, Barking, Essex, England

APPLIED SCIENCE PUBLISHERS, INC.
Englewood, New Jersey 07631, USA

British Library Cataloguing in Publication Data

International Conference on Oil and the
Environment *(1980: University of Edinburgh)*
Onshore impacts of offshore oil.
1. Petroleum in submerged lands — Environmental
aspects — North Sea — Congresses
I. Title   II. Cairns, William J.   III. Rogers,
Patrick M.   IV. Series
338.2   HD9560.5

ISBN 0-85334-974-6

WITH 18 ILLUSTRATIONS

©APPLIED SCIENCE PUBLISHERS LTD 1981

The selection and presentation of material and the opinions expressed in this publication
are the sole responsibility of the authors concerned.

Photoset in Malta by Interprint Limited
Printed in Great Britain by Galliard (Printers) Ltd, Great Yarmouth

Acknowledgements

Advisory Committee

A. M. Dunnett, LLD (Chairman)
Director, Thomson Scottish Petroleum Limited

Professor Sir Kenneth Alexander
Chairman, Highlands and Islands Development Board

Dr J. Morton Boyd (in his personal capacity)
Director Scotland, Nature Conservancy Council

W. J. Cairns
Senior Partner, W. J. Cairns and Partners

J. P. Cameron
Assistant General Manager, International Division, Bank of Scotland

T. F. Gaskell
Executive Secretary Oil Industry International, Exploration and Production Forum

Professor P. Johnson-Marshall
Department of Urban and Regional Planning, Edinburgh University

Professor T. D. Patten
Vice Principal, Heriot-Watt University

Dr Kjell Stenstadvold
Senior Lecturer, Norwegian School of Economics and Business Administration

George Williams, LLD
Director General, United Kingdom Offshore Operators Association

v

Sponsors

Bank of Scotland
European Economic Community (Environment Directorate)
Institute of Petroleum
The Landscape Institute
Norwegian Petroleum Society
Royal Bank of Scotland
Royal Institute of British Architects
Royal Town Planning Institute
Town and Country Planning Association

Special Mention

The organisers wish to thank the Scottish Office and the Edinburgh District Council for their kind cooperation.

Occidental of Great Britain Inc. and the British National Oil Corporation generously made financial contributions to the conference.

This publication is supported by contributions from the British Petroleum Company Limited and from Cluff Oil Limited.

Foreword

It is an honour to have been invited to write a foreword to the proceedings of this timely conference. During this decade, we will witness the height of man's exploitation of the earth's hydrocarbons. As we move into the downward sloping half of the oil production bell curve, even more costly reserves will be wrestled from even more hostile frontiers. The oil industry will, at great expense, overcome tropical swamps, deeper seas, muskeg, icebergs and pack ice.

Dwindling supplies and higher prices ensure these high costs will be accompanied by attractive returns. But what about environmental costs? How much pressure can sensitive frontier environments bear without irreversible damage? How much can industry spend to ease these pressures? How can these outlays best be allocated? This conference proves the oil and gas industry ready to respond to these issues, to search for the real and sensible trade-offs between the urgent need for energy and the need to minimise environmental impacts.

The oil industry, with its privileged combination of capital, revenue and technology, is in a position to prove that growth and progress in the energy sector can continue without damaging delicate social and natural environments. Indeed, this conference signals an awareness that greater cooperation between disciplines and institutions will be needed to cope with the complex choices facing the industry now and in the future.

Energy, environment and economic growth are inextricably linked. At this watershed in man's industrial history, oil capital provides the path to a clean and secure energy future. If the participants of this conference recognise this deeper truth—oil, like the environment, is a precious part of man's heritage upon this planet—the awesome technological challenge of discovery need not be increased by haste, and the bounty of the oil genie need not destroy those who possess the magic lamp.

MAURICE F. STRONG
Chairman,
International Energy Development Corporation,
Geneva

Preface

The depletion of land-based oil reservoirs has led to acceleration in the search for oil in continental shelf structures throughout the world. Discovery of oil and gas in the North Sea led to an exploration and development program of unprecedented intensity, throughout which pressures mounted on the oil industry and environment alike. The organisers of the conference whose preceedings are reported herein shared the view that the lessons learned and experience gained should be internationally reviewed. From this perspective they hoped might emerge new solutions applicable not only to the North Sea and its coastal lands but to other countries facing similar problems. In fact all those who contributed to the conference have helped identify vitally important lessons to be learned from the North Sea experience, which will be ignored at the peril of those undergoing intensive industrial development elsewhere.

In the interests of fidelity we have chosen to retain the order of papers as they were presented at the conference; we hope this will not now appear arbitrary to those who were unable to be present. In material from various countries we inevitably encountered differences between American and British spelling. In view of the spirit of reconciliation between conservationists, socio-economists and oil men in which the conference ended we have therefore followed the enlightening 'blow for liberty from the tyranny of arbitrary custom' struck by Max Nicholson when he wrote *The Environmental Revolution.*

WILLIAM J. CAIRNS
PATRICK M. ROGERS

ix

Contents

xi

Section IV. HAZARDS AND CONTROL

Section V. ISSUES, LESSONS AND CONCLUSIONS

List of Contributors

DR J. M. BOYD, DSc, FRSE, *Director (Scotland), Nature Conservancy Council, Scottish Headquarters, 12 Hope Terrace, Edinburgh EH9 2AS, UK.*

DR R. A. COX, *Technica Ltd, 11 John Street, London WC1N 2EB, UK. Formerly of Cremer & Warner, London.*

MR F. E. DEAN, *Chief Environmental Planning Officer, Production and Supply Division, British Gas Corporation, 59 Bryanston Street, Marble Arch, London W1A 2AZ, UK.*

DR K. E. EGELAND, *Rogalandsforskning (Rogaland Research Institute), P.O. Box 2503, Ullandhaug, N-4001 Stavanger, Norway.*

PROFESSOR J. A. FEHILY, *Fehily Associates (Architects, Town Planners, Landscape Architects), 7 Clyde Road, Dublin 4, Ireland.*

MR J. M. FENWICK, *Manager, Community Relations, British National Oil Corporation, 150 Vincent Street, Glasgow G2 5LT, UK. Formerly Director of Planning, Shetland Islands Council.*

PROFESSOR D. W. FISCHER, *Research Professor, The Institute of Industrial Economics, Breiviken 2, N-5000 Bergen, Norway.*

MR G. GJERDE, *Head of Division of Planning and Negotiations, The Royal Ministry of Petroleum and Energy, P.O. Box 8148 Dep, Tollbugaten 31, Oslo 1, Norway.*

MR R. GOODIER, *Head of Survey and Advisory Section (Scotland), Nature Conservancy Council, Scottish Headquarters, 12 Hope Terrace, Edinburgh EH9 2AS, UK.*

MR H. A. GRAEME LAPSLEY, *Chief Executive, Orkney Islands Council, Council Offices, Kirkwall, Orkney KW15 1NY, UK.*

MRS A. HARRISON, *Regional Administrator, United States Environmental Protection Agency, Region VI, 1201 Elm Street, Dallas, Texas 75270, USA.*

PROFESSOR P. JOHNSON-MARSHALL, *Department of Urban Design and Regional Planning, University of Edinburgh, 20 Chambers Street, Edinburgh EH1 1JZ, UK.*

DR C. S. JOHNSTON, *Director, Institute of Offshore Engineering, Heriot-Watt University, Research Park, Riccarton, Edinburgh EH14 4AS, UK.*

MR F. G. LARMINIE, OBE, *General Manager, Environmental Control Centre, British Petroleum Company Limited, Britannic House, Moor Lane, London EC2Y 9BU, UK.*

MR T. LIND, *Senior Advisor, The Royal Ministry of Petroleum and Energy, P.O. Box 8148 Dep, Tollbugaten 31, Oslo 1, Norway.*

DR M. M. LINNING, CBE, *Consultant, 32A Fountainhill Road, Aberdeen AB2 4DT, UK. Formerly General Manager, Exploration & Production, BP Petroleum Development Ltd.*

MR W. D. C. LYDDON, *Chief Planning Officer, Scottish Development Department, New St Andrew's House, Edinburgh EH1 3SZ, UK.*

MR F. H. MANN, *Manager, Petroleum Engineering, Bank of Scotland, International Division, P.O. Box 10, 38 St Andrew Square, Edinburgh EH2 2YR, UK.*

MR R. MARSTRANDER, *Director, State Pollution Control Authority, P.O. Box 8100, Lorenveien 57, Oslo 5, Norway.*

MR J. McCARTHY, *Deputy Director (Scotland), Nature Conservancy Council, Scottish Headquarters, 12 Hope Terrace, Edinburgh EH9 2AS, UK.*

PROFESSOR I. L. McHARG, *Chairman, Department of Landscape Architecture and Regional Planning, Graduate School of Fine Arts CJ, University of Pennsylvania, Philadelphia 19104, USA.*

DR E. M. NICHOLSON, CB, CVO, LLD, *13 Upper Cheyne Row, London SW3 5JW, UK. Principal, Land Use Consultants, London. Formerly Director General of the Nature Conservancy.*

DR G. NORCROSS, *Manager, Environmental and Safety Affairs, ICI Europa Ltd, Everslaan 45, B-3078 Everberg, Belgium. President, International Association of Environmental Coordinators, Belgium.*

DR C. G. RAMSAY, *Cremer & Warner Scotland, 222 Union Street, Aberdeen AB1 1TL, UK.*

MR M. SARGENT, *Director, Department of Planning and Development, Orkney Islands Council, Council Offices, Kirkwall, Orkney KW15 1NY, UK.*

DR D. H. SLATER, *Technica Ltd, 11 John Street, London WC1N 2EB. Formerly of Cremer & Warner, London.*

MR T. F. SPROTT, *Director of Physical Planning, Grampian Regional Council, Woodhill House, Ashgrove Road West, Aberdeen AB9 2LU, UK.*

DR K. STENSTADVOLD, *Senior Lecturer, Norwegian School of Economics and Business Administration, NHH, Helleveien 30, N-5000 Bergen, Norway.*

MR C. STUFFMANN, *Head, General Planning and Improvement, Environment and Consumer Protection Service, Commission of the European Communities, Rue de la Loi 200, B-1049 Brussels, Belgium.*

DR J. F. TAYLOR, *Shell UK Exploration and Production, Shell-Mex House, Strand, London WC2R ODX, UK.*

Introduction

By the Rt. Hon. George Younger MP, Secretary of State for Scotland

We in Scotland have over the past decade experienced an astonishing upsurge in activity offshore and onshore, through the exploration and exploitation of hydrocarbon resources in the North Sea, and the impact on communities in the areas affected has necessarily been considerable. Substantial social and physical changes, through technological development, are of course no new experience for the Scots who were early developers in major industries such as steel and shipbuilding and also in the shale oil business.

What was necessary—sometimes in parallel with offshore activity but often in advance of it—was the planning of and considerable investment in related facilities and infrastructure. On the former we have seen throughout the 1970s the construction of oil terminals at Sullom Voe in Shetland—destined to be the largest in Europe—Flotta in Orkney, Nigg Bay in the Cromarty Firth and Hound Point in the Firth of Forth. A major gas treatment plant has been built at St Fergus and a number of platform construction yards established across the country. Such yards and allied oil industry requirements have provided a chance for the Scots to show their adaptability in the use of their long-established skills in heavy engineering and shipbuilding. On infrastructure major investments have been made in service bases, airport facilities and roads, and there has been extensive expenditure also on housing and similar social requirements.

Given the scale and rapid growth of this activity it was inevitable that events would not always occur as predicted. The pace is such that decisions have to be taken very quickly since a chance missed can be a chance lost. In my view, however, all concerned—in government, local authorities, oil companies and other public bodies involved in these developments—can look with pride at what was achieved, the speed with which it was accomplished and the regard which at the same time was paid to the environmental, social and cultural implications.

Through all these joint experiences and efforts many lessons have been learned and it was therefore very fitting to bring together in Scotland, at the International Conference on Oil and the Environment in September 1980, those who had been closely involved in developments in the North Sea, in the UK, Norway and elsewhere, in order to discuss and compare experiences and to demonstrate that the interests of the developer, the controlling authorities and the environmental agencies can be reconciled.

For the same reasons I welcome the decision to publish this record of the proceedings of the conference. I am sure it will prove a very valuable source of information and guidance, not only for those currently involved in North Sea developments but for those who find themselves confronted with the prospects of similar developments elsewhere.

Section I

NORTH SEA COASTAL LANDS

Chairmen

Papers 1 and 2
R. Keir Watson Esq., Chairman, British Polar Engines,
Glasgow and London, UK

Papers 3 to 5
Dr Jean Balfour, CBE, Chairman, Countryside Commission
for Scotland, UK

1

Environmental Aspects of the Development of North Sea Oil

F. G. LARMINIE

*Environmental Control Centre, British Petroleum Company Limited,
London, UK*

ABSTRACT

*Possible conflicts between oil- and gas-related development and other existing
uses of the sea, seabed, coastline and contiguous North Sea lands are sum-
marised. Generally, environmental standards achieved by the North Sea oil
and gas industry have been high and environmental considerations are, like
safety, becoming an integral part of modern oil company management.*

INTRODUCTION

Development of North Sea oil and gas resources could have environmental consequences in any or all of the three main physical divisions of
the basin and its hinterland, namely

   (i) the sea and the seabed—the basin floor and its contents;
  (ii) the coastline—the basin rim or margin;
 (iii) the adjacent (or contiguous) lands.

In addition there are possible effects on society, and while man is an
integral part of ecosystems this less well defined impact is regrettably
isolated under the heading 'socio-economic effects'. Here the reader is
cautioned to beware of the oft repeated catch-phrase 'the impact of man's
activity on the ecosystem'. It would be difficult, if not impossible, to find
an ecosystem that has not been modified by man's activities, and the all

5

too common practice of considering man as a sort of *deus ex machina* apart from the dynamics of ecosystems is totally fallacious. This could easily become the subject of a whole paper but for the present it is sufficient to sound a warning and revert to the subject matter, namely the environmental factors which should be considered in any offshore development.

SEA AND SEABED

Existing uses include fisheries, shipping, gravel extraction (applicable only to the predominantly gas province of the shallower waters of the southern North Sea) and recreation (which mainly applies to inshore coastal waters).

Possible conflicts between existing uses and the offshore oil industry include

  (i) location of platforms and other production installations;
 (ii) location of sub-sea pipelines;
(iii) seabird migration—flares on platforms,
                    —annual auk migration across the
                      North Sea, e.g. from Shetland to
                      Norway; the birds are flightless
                      and swim across, and would be
                      extremely vulnerable in the event
                      of an oil spill on or near the
                      migration route;
 (iv) pollution arising from an accident to a platform, pipeline or other
      production installation, or a blow-out in an exploration or pro-
      duction well.

At sea such pollution could affect commercial fisheries, particularly nursery grounds, while inshore it could have serious effects on the coastline and in shallow coastal waters. Severe pollution could result in a heavy mortality of planktonic, benthic and littoral fauna and flora. The benthos are usually less severely affected than the others because they are less exposed to oil. However, the changes caused by pollution have to be measured against a widely fluctuating natural background and while there may be no doubt that an oil spillage or the effluent from a process plant is directly responsible for observed changes in the local biota, there are serious pitfalls in attributing all such changes to a pollution incident in the absence of pre-development data.

While oil pollution may kill organisms in the sea and on the shore and may result in dramatic ecological change, usually the mortality is without long-term consequences and the ecological change is within the normal range of fluctuation in marine environments. Indeed, measurement of the rate of recovery of an ecosystem that has been subject to pollution can be difficult, if not impossible, without adequate knowledge of the natural variations in the marine environment.

Rigorous appraisal of the potential impact of the project will reveal gaps in our knowledge and it may be necessary to commission research to get the necessary data on unmodified ecosystems. The migration of fledgling and flightless adult auks which swim from Shetland to Norway annually or, by analogy with the introduction of lighthouses, the possibility that platform flares might affect migrating birds are just two examples from ornithology (where, sad to say, there is still far too much emotion and not nearly enough science).

It is interesting to note that a program of bird observations from platforms has been instituted and with professional guidance operating staff are being trained to make scientific observations of birds on and about the platforms. Indeed offshore development in the North Sea has brought about a great increase in our knowledge of the physical characteristics of the sea, the seabed, the coastline, and the associated flora and fauna. In meteorology long- and short-term forecasting has benefitted, particularly with respect to weather windows at sites for vital construction operations. In oceanography we have much better knowledge of bottom currents, sea water composition and physical properties (including temperature and salinity) and wave propagation and characteristics (e.g. height, length, period, frequency).

The availability of this information from the open sea results from the refinement of having what are, in effect, instrumented observation platforms and this is an important point when one thinks how subjective much of the data was 15 years ago, before the search for oil in the northern North Sea began.

THE COASTLINE

Existing uses of the coastline include residential use, recreation (beaches, marinas, coastal golf courses), commerce (ports and harbors and their associated industries), nature reserves (or SSSI's—see McCarthy, these proceedings—or National Trust heritage properties), defense (naval bases, airfields, etc.) and agriculture.

Possible conflicts with those uses include

(i) sensitive shorelines which comprise important coastal habitats could be affected by (a) pipeline landfalls, (b) aircraft movements: the potential impact of helicopter overflights on seabird colonies, particularly in the breeding season, was recognised early on. A series of experimental overflights established the optimum altitude for crossing coastal seabird colonies;

(ii) amenities are discussed elsewhere in these proceedings. However, it is worth mentioning that the Forties pipeline came ashore at the south end of the Cruden Bay golf course. A stipulation was that not a cup of sand from the construction should contaminate the adjacent green. Rumor has it that once the protective sheet-piling was in place, such was the dedication of the golfers that they were unaware that history was being made beyond the 9th green as the first oil pipeline from the British sector of the North Sea came ashore (see Linning and Larminie, these proceedings);

(iii) disruption of traditional activities in harbors, e.g. competition for berthing space, hard standing, etc., which could displace the shore facilities associated with traditional seafaring activities;

(iv) pollution from pipeline failure or from terminals, which can be accidental or chronic. The latter will include both liquid effluent and emissions to atmosphere from processing operations.

Examples of remedial action to minimise the impact of pipeline land-falls include the Cruden Bay (Forties) and Grut Wick, Shetland (Ninian), landfalls. The long curved sandy beach at Cruden Bay at the low energy end of the geomorphological spectrum is in sharp contrast to the rocky, high energy cobble/boulder beach at Grut Wick. Both were successfully restored and it would take a practised eye to detect the landfalls, and there is certainly nothing to show the average holidaymaker where the lines come ashore. They have even eluded some of our own public-relations men bent on demonstrating it to visitors.

THE CONTIGUOUS LANDS

The existing uses are well known, and just to list them would be redolent of the contents page of a geography text.

Possible conflicts arise from

(i) terminal construction or platform construction yards;
(ii) esthetic considerations;
(iii) creation of imbalance in stable human communities through (a) massive influx of labor—in the minds of some, the latter-day equivalent of the Viking hoardes, (b) poaching elements of the existing labor force away from their traditional activities;
(iv) pollution—once again a similar catalogue of accidental spillage, discharge of liquid effluents, emissions to atmosphere with the additional factor of noise.

The above incomplete catalogue of possible conflicts/impacts raises the question of how best to identify and deal with the environmental issues germane to all major developments, not only those offshore in the North Sea. This question is considered in various papers in these proceedings and is not, therefore, dealt with in detail here.

Environmental impact analysis is not, of course, obligatory in the UK, but in BP it is carried out for all major capital projects. It provides part of the material for making decisions through rational debate by reasonable men which is a hallmark of the UK planning process. It is a process of cooperation rather than confrontation which latter process regrettably characterises the US approach to similar projects and leaves the field wide open to legal skirmishing, and as an ecologist who shall be nameless once observed in a moment of frustration 'In the Federal Ecosystem, are they scavengers or predators?'.

However, to acquire a dynamic dimension, impact analysis must be linked to a program of monitoring of a project before and during construction, and during operation. Only thus can one attempt to reconcile actual versus predicted impact, thereby gaining fresh insights to be applied in the design, construction and operation of future projects.

INDUSTRY'S PERFORMANCE RECORD

It can be fairly claimed that industry's performance record at sea has been very good and there have been no blow-outs in producing fields in the UK sector. Inshore at terminals, the record was marred by a serious incident at Sullom Voe shortly after start-up. However, major crude-handling terminals are not new in Scotland, and virtually all the oil for Scotland came in through the Finnart terminal on Loch Long before the

North Sea oil was ever discovered. Because of its fine record and as it is an example of a terminal located in a deep narrow fjord capable of handling very large tankers, it was used as an analogue for the proposed Valdez terminal of the Trans-Alaska pipeline, and as such was inspected by the fishermen from Cordova, Alaska, as part of their efforts to not only assess, but visualise the operations of a very large crude oil terminal in a long narrow fjord.

The socio-economic effect of North Sea oil development is difficult to assess. At the end of the day will it be great? In the macro-economic sense of the contribution towards the nation's self-sufficiency in oil and its balance of payments the answer is obvious, but at a local or regional scale it is less so. At one time the potential impact was spoken of in the same terms as the Highland clearances but I think that even the author of that notable polemic *The Cheviot, the Stag and the Black Black Oil*[1] would admit that the consequences have not been nearly as dire as they imagined 6 or 7 years ago. That there has been some disruption and change is undeniable, but change is by definition neither good nor bad and its perceived direction is very much dependent upon the perception and attitude of the beholder.

In the broadest sense I believe that we have developed a sound approach to environmental conservation in a difficult area based on good science and sound engineering and that, like safety, environmental considerations are well on the way to becoming an integral part of the management of a modern oil company's affairs. That is now demonstrably the case in certain companies. But it is no grounds for complacency and it is incumbent on all of us to make every effort to ensure that at the very least we maintain the high standards which I believe have characterised the development of the offshore oil and gas industry in the UK sector of the North Sea.

REFERENCE

1. McGrath, J. (1974). *The cheviot, the stag and the black black oil.* West Highland Press, Kyleakin, UK.

2
The North Sea and its Coastal Lands

E. M. NICHOLSON

13 Upper Cheyne Row, London, UK

ABSTRACT

The North Sea and its coastal lands comprise a single complex interacting system, the stability of which is currently at risk. The problem of handling it is aggravated by approaching it piecemeal, in terms of unrelated or overlapping disciplines, policies, priorities and jurisdictions. The inadequacy of such an approach may be more apparent when the subject is considered from an environmental standpoint than from others.

Many landward processes and activities are shown to affect the North Sea, and vice versa. Such interactions are dynamic and subject to far-reaching and rapid change. Failure to foresee them can lead to heavy losses. There is an immediate need for concerted remedial and precautionary measures, and while these should be adaptable and should not avoidably conflict with global approaches under consideration, they are too urgent to be deferred until agreed solutions are found for the oceans generally.

INTRODUCTION

It ought to be a truism, determined both by ecology and economics, that a sea is inseparable from its coastal lands. In fact, the entire structure of government, as well as that of most scientific and professional disciplines, conspires to mislead public opinion into treating the problems involved in bits and pieces. The case of the North Sea is especially unfortunate. Few seas are as rich in resources calling for concerted study and management. Few seas have been surveyed, at least fragmentarily, for so long or in so much scattered detail. No important sea is more compact, or faces such a closely akin and culturally related group of fringing nations. Yet it is arguable that the North Sea peoples had a better awareness of their common links and heritage in the Dark Ages, under the Angles, Saxons and Vikings, than they have today, even if that awareness more often led to bloodshed.

It is true that, from the international stratospheric plane of NATO and the EEC to numerous *ad hoc* bodies and national agencies, North Sea matters are often on the agenda. The record of action, and even of coordinated research and planning is, however, clearly and lamentably inadequate. We are not yet in sight of being able to retrieve from a computer what is known about the North Sea and how its pressing problems interrelate. It is still given to few to grasp any substantial part of what is more or less secretly known in one cloistered institution or department or another and who, if anyone, is supposed to be interpreting and disseminating it, let alone doing anything practical about it. There is, of course, plenty of action, but it is all inevitably based on partial understanding, and too often on guesswork. Environmentalists and ecologists worry about this, but few others seem to. This paper seeks to piece together some of the obvious implications of a number of discrete specialist studies and findings, and to stimulate interest towards the balanced and comprehensive understanding and use of the North Sea by those concerned.

THE NORTH SEA ECOLOGICAL SYSTEM

Jacques Piccard has asserted that the North Sea begins in Switzerland, as a dweller up the Elbe could claim that it begins in Czechoslovakia, or a Highlander that it springs in both directions from the Pass of Drumochter. It is true that the composition and quality of North Sea waters is markedly influenced by the input of fresh water from its catchments, and by the burden of sediments, nutrients and pollutants contributed by their river systems. Distribution maps of salinity ratios, fisheries and of such elements as zinc, manganese, nitrate and phosphate[1] indicate the extent to which the land influences the quality and potential of the sea.

The narrow zone which is neither wholly land nor wholly sea is of outstanding significance for both. Human settlements, such as ports and seaside resorts, which have taken advantage of some of the resultant resources have neglected or injured others. Waterborne traffic by rivers has also depleted or damaged them, not only as themselves but as vital feeders and complements of the sea to which they belong. Nowhere are such injuries more painfully evident than along the Rhine, although the cost and effort demanded in cleaning up the Thames, the Tyne and other North Sea rivers after much more limited pollution is also a measure of the problem.

In order to discern the more complex and subtle interrelations between land and sea it is convenient to focus on the coastal and hinterland plains below the 100 m contour. In some cases they extend inland some 200 km; in others they narrow to a strip or a mere rim above high water mark; at some points, especially in Norway and northern Scotland, they are squeezed out almost entirely between the mountains and tidewater. In their upper zones they enjoy a milder climate and easier topography than the continental hinterland and they play a substantial role in making the climate of much of western Europe more Atlantic than typically continental. Strategically and economically they also ensure, through communications and politics, that the hinterland looks to the coast rather than vice versa as in Spain or Russia.

The coincidence that western capitalist expansion and the mastery of marine navigation occurred simultaneously with widespread and prolonged setbacks to the political and economic development of continental Europe and with the opening up of rich opportunities in the Americas and the Orient has imposed an enduring distortion on the development of the North Sea lands, which is only gradually being modified in the post-colonial age. This is apparent both in the chronic depression now experienced by Merseyside and Clydeside, and the recent eclipse of the Port of London by Rotterdam-Europoort and Antwerp exploiting the potential of their rich hinterland. The growth of such British ports as Teesmouth and Felixstowe has been on a much more modest scale, but it is clear that, leaving oil out of account, there is a strong trend towards economic growth and a shift of population centering on the North Sea.

One facet of this has been the series of large-scale projects involving particularly the lower North Sea coasts, of which the reclamation of the Waddenzee, strongly resisted by environmentalists, the Dollart project on the Ems and the proposals for artificial islands off The Netherlands and elsewhere, have not so far been realised, in common with the Channel tunnel, the much-discussed Maplin airport and seaport off Essex, and the large-scale reclamation of the Wash.

Ecologists are particularly concerned over the widespread illusion that mudflats, salt marshes and other coastal areas forming attractive targets for reclamation are unproductive. In fact biologically they rank high in productivity and, as nurseries and refuges, they play a large part in sustaining commercial fisheries as well as wildlife over much wider regions. It is true that, if reclaimed and reconditioned, such areas also tend to be of high agricultural productivity. So do flat lowlands, sometimes even beneath sea level, which have been subject to seasonal flooding over centuries or millennia, such as the fens of east England. There is, however, a price to be paid not only in wildlife and fisheries, vegetation and environmental values, but in impoverishment of the heritage of the gene pools.

There is also a time factor. Wetlands, including salt marshes, sand dunes and other coastal features, may be subject to natural change but, in one form or another, they continue over centuries and millennia. They can, however, be blotted out at short notice by nuclear power stations, onshore installations, steel mills, hotels and other structures which can have a useful life of only a handful of decades or even less, and which will sterilise their sites, if not forever at any rate for a very long time. The instability, impulsiveness and undependability of current technology and its commitments is a source of growing concern to environmentalists. They see irreplaceable ecological assets, derived from many millennia, being blotted out with reckless haste to accommodate allegedly essential developments which are then often seen to have been misconceived even on their own narrow and short-sighted assumptions.

It is true that the North Sea itself is a highly unstable phenomenon. At its southern end, coasts are sinking with remarkable rapidity, demanding costly and sophisticated projects such as the still incomplete Silvertown Barrage across the Thames. Here it is not a question of reclaiming more land from the sea, but of checking as long as possible the sea's attempt to reclaim more sea from the land. That menace is aggravated by the tendency in certain weather conditions to develop a storm surge. This, when it coincides with high tide and flooded rivers, can bring instantly to a crisis the ever present threat of inundation.

The opening of the southern exit of the North Sea opposite Dover is so geologically recent that prehistoric man could still walk across to the future island of Great Britain, beside rivers whose outlets were later deflected by the new tidal regime. Towns and villages have been washed away, while elsewhere new land has emerged, as at Tentsmuir by the mouth of the Tay, where the acreage of the National Nature Reserve, declared for the Nature Conservancy in 1954, has constantly had to be revised upwards. Exactly what happens on the seabed remains obscure but there is clear evidence that the reliability of submarine soundings on charts can be vitiated by shifting sands. Man-made changes have yet to reach a scale comparable with those long inherent in natural processes.

The range and influence of these interactions between the North Sea and its neighboring lands have lately been intensified by the finding and exploitation of undersea oil and gas resources which have transformed the geography of west European energy. The localised impact, however, has so far been concentrated upon rather few centers, either pre-existing such as Aberdeen, Stavanger and Teesmouth or newly developed such as Sullom Voe, Flotta and St Fergus. With such exceptions the socio-economic repercussions have been diffuse and widespread, for example through the effects on other industries of higher exchange rates and insulation from some of the pressures of the world energy crisis.

The coincidence of acute depression in the fishing industry, largely through neglect of urgently needed conservation measures during the 1960s and 1970s, has created pockets of depression in a series of port areas, some of which have been partly compensated by oil revenues while others have not. On the environmental level both abrupt expansion and abrupt contraction bring serious problems, which are complicated by recent structural shifts towards regional and long-term economic and physical planning. The effect of these shifts can be to spread risks and broaden the base of available expertise and help, but it can also remove decision-making to points more remote from the problem, and delay action while hitherto unconcerned hinterland interests try to adjust to novel seaborne impacts on their affairs. We are witnessing a blurring of the previously sharp lines between coastal and inland communities but the process is retarded and confused, in the continuing absence of informed public opinion on both sides of the North Sea, over the implications of being a modern North Sea nation.

In a world perspective there is growing interest in the oceans and seas from many angles. Regional seas are coming under close attention especially in such leading cases as the Mediterranean and the Caribbean, with which the North Sea should rank. Indeed, but for the transatlantic flow of Caribbean currents around Scotland and through the Strait of Dover, the North Sea might have missed being ice-free, and its surrounding lands might be boreal rather than temperate. There is no time here to make comparisons which would demonstrate what a rare combination of good fortune is enjoyed by those who dwell around the North Sea, even if only the hardiest of them can be happy swimming in it.

Around its great U-shape, its basal funnel between the Low Countries and south-east England (and still more if we include its inverted V extension through the Skagerrak to the Kattegat), the North Sea displays a remarkable range of habitats. These include arctic–alpine heaths, deep long fjords, low moorlands, high mountains, islands and skerries, bird-cliffs, sand dunes, salt marshes, firths, torrents and placid estuaries, vast mudflats, fertile farmland, tidal rocks, sandbanks and shingle spits, shallow fishing banks and deep trenches, with seabeds of mud and sand, gravel and rock, peppered with old wrecks and striated with modern pipelines and cables.

In a visual age it is rich in photogenic features. In an exploratory age it offers many varied opportunities for following up with modern techniques the notable pioneer work which has been done on it, especially earlier in this century. In an age of conservation and development of natural resources it presents great challenges which have hitherto been only dimly perceived and fragmentarily taken up. In an age of growing European self-awareness it offers one of the most magnificent opportunities to focus and crystallise common traditions and common heritage in a vigorous and vivid new pattern of cooperation. In an age of narrowing horizons and diminishing opportunities it still presents some worthwhile openings for fresh developments.

Why, then, is the outstanding achievement of the oil and gas technologists and managers, who have performed miracles during the past decade or so, not better matched by a broad and lively renaissance of North Sea life? Responsibility must be very widely shared. Once again our modern intelligentsia and the intellectual and artistic establishment have almost totally failed to understand and follow up the challenge of western Europe's new frontier. Only in The Netherlands is there any sign of real awareness of the European cultural and political significance of the North Sea. Even where interest has been aroused in so relevant an aspect as the Viking age, there has been no attempt to relate it to the problem of renewal of the common North Sea heritage. The early death of Benjamin Britten of Aldeburgh has robbed us of the only outstanding champion of North Sea traditions in our cultural life.

Although nearly all the North Sea countries were on the same side in the World War II, and close ties existed between, for example, Britain and Norway, The Netherlands and Belgium, these have not been fully maintained. The late British entry into the European Community, and the subsequent disagreements, have helped to block the way for Community leadership to promote closer relations around the North Sea, while the later adhesion of Denmark but not of Norway made a further obstacle. Rivalry between ports and fishing interests, and some sections of agriculture, loom larger than such common enterprises as the Anglo–Dutch Unilever and Royal Dutch-Shell multinationals and the more recent Anglo–Norwegian oil development. On the German side cordial relations with Hamburg and Bremen have not been able to alter the fact that transfrontier links are naturally viewed as of more substance than hands across the sea. While there appear to be plenty of promising prospects for complementary enterprise, such successful examples as the Dutch participation in the English bulb industry are still few and rather small. Progress in adding to them could do much to provide a firmer economic base for the creation of a true North Sea community. There is still an abundance of mutual goodwill and ready understanding, and nowhere else in Europe do so many people speak and read English really well as across the North Sea.

In environmental matters links have long been close and practical. In ornithology the former British colony of Helgoland has exercised a profound influence on bird migration studies, not least in Scotland. Of my fellow editors of the Oxford seven-volume *Handbook of the Birds of Europe, the Middle East and North Africa*,[2] three are Dutch and the rest British.

Nevertheless, while the human potential for a much more important North Sea link-up undoubtedly exists, the necessary inspiration and impetus to create it evidently do not. To some extent this may be due to the exaggerated pre-occupation of British opinion, first with the Empire and then with superpower politics, and with the adverse course of European Community evolution. Perhaps to an even greater extent it is a lack of realistic thinking and leadership concerning the long-term interests and potential of Britain and also of her North Sea neighbors, although here The Netherlands is a shining example.

CONCLUSIONS

Whatever may have gone wrong it is not too late to start remedying it, and indeed, with the closing of many options which looked more tempting earlier, the prospects may now be better. There are four areas in which sober and well conceived initiatives may now be effective.

First, at a scientific and technical level, the urgent need is to assemble, digest and present in a clear, integrated, up-to-date form the factual, statistical and cartographic information which exists either as raw material or in specialised accounts in the many unrelated institutions and services concerned with one part or another of the whole. This should be focused in a major International Atlas of the North Sea, under appropriate sponsorship, to be kept up-to-date with ancillary textual, statistical and cartographic reviews. Under the auspices of the Royal Geographical Society in London, preliminary inquiries and talks on this have been held in this country, The Netherlands, the Federal German Republic, Norway, Denmark and elsewhere. The desirability of such a project is widely and keenly recognised, and the Geography Department of the London School of Economics is ready to undertake the task in full consultation with the many institutes and organisations concerned. What they most urgently need is money to start the work. It is something which needs to be done quickly and well.

Secondly, the time seems to be ripe for developing a concerted program of education and enlightenment about the significance of the North Sea. Some of the obvious points and areas to be covered have already been mentioned. The spearheads of such a campaign should clearly be television and radio services all around the North Sea coast, which could make and exchange film and tape material on a large scale and on coordinated lines, and the coast universities, from Aberdeen, Dundee, St Andrews and Edinburgh to Newcastle and Durham, East Anglia, Colchester, Ghent, Delft, Leiden, Hamburg to Aarhus, Copenhagen, Lund and Oslo, with others interested, including London. Could not a few leading academics circulate preliminary proposals to their colleagues, to be followed by some kind of meeting to arrange exchanges of lectures, visits and publications designed to promote fuller and deeper understanding of the main elements in their common problems and their potential contributions?

A third practical initiative would be the early establishment of a non-governmental North Sea Forum for the study and discussion of new approaches to the integrated management of the North Sea. A resolution to this effect was adopted at the Hague by The Netherlands North Sea Working Group,[3] which asked the Commission of the European Communities to consider assisting the project, financially and otherwise. This request has apparently met with no response, and since the Greenwich Forum[4] and other influential meetings have managed without receiving government or international funds, it seems time to go ahead on some alternative basis. As was agreed at the Hague (op.cit), such a Forum should not be limited to EEC countries.

It would not in any way remove the need for planning and agreeing upon the form of an intergovernmental agency to create and supervise an organ or organs for the coordinated management of North Sea activities and projects, exclusively at the level where coordination is essential, and for the conservation of its natural environment and renewable resources. Part of this field is already covered intergovernmentally by the official international group originally convened by the Nature Conservancy Council which met again in Norway in March 1980. It would seem appropriate for the European parliament to take the initiative on this or, failing it, the Council of Europe.

It is sometimes urged that such measures for particular seas should be shelved, pending the resolution of global and seemingly interminable discussions about the law of the sea and kindred problems. This is a misunderstanding. The kind of scientific and operational matters here indicated can and must be sorted out on their regional merits on the spot, and this in no way prejudices the juridical and cosmic negotiations, the main obstacles to the success of which have been imported by those taking part in them.

REFERENCES

1. Lee, A. J. & Ramster, J. W. (1977). *Atlas of the seas around the British Isles.* Fisheries Research Technical Report, No. 20. Ministry of Agriculture, Fisheries and Food; Directorate of Fisheries Research, Lowestoft, UK.
2. Cramp, S. (Ed.) (1977). *Handbook of the birds of Europe, the Middle East and North Africa.* 7 vols. Oxford University Press, UK.
3. North Sea Working Group (1979). *Seminar on the North Sea and its environment; uses and conflicts.* P.O. Box 90, 8860 AB Harlingen, The Netherlands.
4. Greenwich Forum (1979). *Europe and the sea; the case for and against a new international regime for the North Sea and its approaches.* International Conference, Greenwich, UK. (See also Sibthorp, 1979.)

BIBLIOGRAPHY

Dietrick, G. & Ulrich, J. (1968). *Atlas zur Ozeanographie.* Bibliographisches Institut, Mannheim, West Germany.

Goldberg, E. D. (Ed.) (1974). *North Sea Science.* M.I.T. Press, Cambridge, Mass., USA.

Kent County Council, Planning Department (1980). *International seminar: management of the littoral—Channel and southern North Sea—the next 10 years.* 3 vols. English Version: Public Relations Officer, County Hall, Maidstone, UK.

Nature Conservancy Council (1977). *Report of the London international meeting on wildlife and oil pollution in the North Sea,* NCC, London, UK.

Nature Conservancy Council (1979). *Nature conservation in the marine environment.* Report of the NCC/NERC Joint Working Party on Marine Wildlife Conservation. NCC, London, UK.

Scottish Development Department (1974). *North Sea oil and gas; coastal planning guidelines.* SDD, Edinburgh, UK.

Sibthorp, M.M. (1975). *The North Sea; challenge and opportunity.* Report of a study group of the David Davies Memorial Institute of International Studies.

Sibthorp, M.M. (1979). Viewpoint; the Greenwich Forum. (Summary and conclusions of the Conference listed in ref. 4.) *Marine Pollution Bulletin,* **10,** 186–8.

World Wildlife Fund (1979). Project 1411; The Waddensea, Netherlands, Germany and Denmark, conservation program. *World wildlife yearbook 1978–79,* pp. 26–30. WWF, 1196 Gland, Switzerland.

3

Control Strategies: Physical Planning

W. D. C. LYDDON

Scottish Development Department, Edinburgh, UK

ABSTRACT

Oil and gas developments in Scotland have resulted in significant economic, social and physical changes in the last 10 years, primarily between 1970 and 1975: 60 000 jobs created; 2000 hectares of land developed; the lifestyle of 14 communities affected. Such developments have special characteristics: diverse nature of projects, dispersed locations, speed and uncertainty of development; the need for collaboration and cooperation; often working near the forefront of technology.

During 1970–75 all five major oil and gas terminals, all service bases, four land pipelines, some 15 platform yards, and 50 other major developments were approved rapidly without a public inquiry. There were four public inquiries for platform yards in environmentally sensitive areas. In many rural areas major industrial projects had not been experienced nor expected, but planning authorities, with consultants, quickly carried out contingency plans and environmental assessments. The Scottish Office surveyed the coast and produced national planning guidelines and advice on environmental assessment. Although criticised, land-use planning and control machinery proved adequate. New legislation was required for licensing refineries, designation of sea areas, funding of land restoration and finance for local authorities.

Possible lessons for physical planning strategies in other countries are: allowing for uncertain effects of exploration; appropriate public bodies for coordinating and communication; adapting the instruments for planning and control at each level of planning; cooperative working to achieve forward planning and environmental assessment; appreciation of the importance of the rate and type of change rather than absolute levels; attempting to deal with worry and stress caused by rapid changes and the instant reaction of the news media.

INTRODUCTION

Oil and gas developments in Scotland have, during the last 10 years, created significant economic, social and physical changes in many areas. Some 60 000–70 000 jobs have been created, 14 individual areas and communities have been involved, and 2000 hectares of land have been developed. Most of these changes and developments happened between 1970 and 1975. History may suggest that these years distinguish what may be called the first phase of land-based oil developments in Scotland. All the main terminals, service bases and construction yards which are in operation today were sited by the end of 1975. The second phase may be considered to have started in 1975 with the proposed development of an oil refinery at Nigg on the north shore of the Cromarty Firth, and an ammonia plant near Peterhead. This phase continues today with the developments at Mossmorran on the north shore of the Forth and with the current consideration of the gas gathering pipelines. This paper is mainly concerned with the first phase, since the land-use planning problems and lessons all arose then, but it is possible only to summarise the main characteristics of and authorities' reaction to land-based developments of those crowded 5 years, from which the conclusions are drawn.

MAIN CHARACTERISTICS OF LAND–BASED DEVELOPMENTS

The term 'oil-related developments' covers a wide range of different forms of land-use requirement and construction schemes. These include such diverse projects as major oil and gas terminals requiring perhaps 200 hectares of land, service bases usually involving development of existing harbors, platform construction yards with their particular siting and marine requirements, land pipeline routes with booster stations, and a wide variety of manufacturing and servicing establishments together with administrative offices. For planning there was a significant difference between the impact of the short-term contract-based operations associated with exploration and construction, and the longer-term location of the terminal, storage and pipeline installations. Some of these projects involved unusual forms of construction, working at the forefront of technology and requiring rapid understanding by planning authorities of the likely impact of each scheme. Others in established urban areas created fewer planning problems.

It is a particular feature of oil development in Scotland that it has not involved concentration of land-based development as has occurred elsewhere in the world, although many operators might have foun .his more convenient. However, the impact would have been more severe if it had all taken place in one community. As it was some 14 different communities were affected, stretching from Shetland in the north-east, generally down the east coast of Scotland, but also affecting the Clyde and other parts of the west coast for platform construction. Many of the places were rural communities in which a major enterprise paying an industrial wage had never been experienced nor expected. While welcome because of the economic uplift that was given to the area and the reversal of depopulation, the effects of these new forms of development on the lifestyle in the area was considerable.

It was some time before it was realised that the wide span of developments was not the sole concern of, nor indeed under the control of, the major oil companies. Many contractors, subcontractors and other servicing operations were involved, each concerned with securing its own land base and being in the position to tender for the work that was going to be available. Many of these operators were inexperienced at setting up developments in Britain under the British planning system and this, in the early days, led to several stresses and strains. In addition many people submitting planning applications were doing it on the chance of gaining a contract, or of speculating in the land. The public were not always able to distinguish between a planning application which was probing the possibilities and one which had some foundation in the main requirements and operations. To them, all proposals were likely to materialise the next day.

Where the oil or gas would be found, the mix of the find, whether it would prove commercial and whether it would be brought ashore were for each particular exploration area uncertain. Many found it difficult to believe the oil companies and indeed the government could not be certain of what was going to happen. Nothing leads to stress and worry more than uncertainty. It was only as the pieces of the jigsaw became more certain and were fitted to the previous bit that the overall pattern gradually began to emerge. Once the commercial judgement had been made, however, time was money and decisions were required quickly. The land-based developments required to make the United Kingdom self-sufficient in oil and to double the throughput of natural gas all occurred between 1970 and 1975. This is normally the period in which it takes a planning authority to draw up and to start to implement a development plan, or the period it takes to plan a new town. In addition to the large number of operators and planning authorities involved there was also a wide range of government agencies and departments who were centrally concerned with various aspects of the operations. This made the need for new forms of collaboration and cooperation imperative. Planners advising authorities approving the schemes had to reach a rapid grasp of the new technologies involved, had quickly to gain information about the environment which was likely to be affected and reach some appreciation of the likely consequences for the surrounding communities. With the uncertainty, speed and diverse nature of the projects it was little wonder that the news media reacted in what now may be seen as an over-excited mood. This resulted in every new development being seen either as a further bonanza, or as adding to the rape of Scotland, or as a boom to be followed by a damaging decline. Planners, as always, had to form a balanced view.

DEVELOPMENTS AND REACTION

Having considered these major characteristics it is now appropriate to review the major developments and how the planning authorities and other agencies handled them. One of the first steps that the Secretary of State, as Planning Minister for Scotland, took was to direct that all major oil-related developments should be referred to him so that he could, if necessary, take the final planning decision. Between 1970 and 1979 there were 70 major applications thus referred. It should be emphasised that all five major oil and gas terminals, all service bases, four land pipelines and some 15 platform yards and 50 other developments were all approved rapidly and without major objections or a public inquiry. During the first stage only four public inquiries occurred and they were for platform yards in environmentally sensitive areas on the west coast.

Such was the need for speed that many planning applications were perforce speculative or the proposition was being elaborated at the same time as the application was lodged, in order to be in a position to offer a tender for the work. At one stage, for example, planning approval had been given for 15 platform sites but only three were in operation. We had 'proposals' for seven oil refineries around the coast of Scotland and not one was in fact built. So in addition to analysing projects which materialised, government departments and planning authorities also had to consider a wide range of planning applications where work did not proceed, e.g. at Dunnet Bay near John O'Groats and Loch Eriboll further west along the northern coast. Reaction to all this activity was immense and local authorities who were at the same time contending with reorganisation have not, perhaps, been given sufficient credit for the way in which they handled oil-related developments. In each oil action-area, development plans had to be drawn up quickly for places where major industrial developments had not been contemplated, and at an early stage development planning or project analysis was being carried out in every one of these areas. Overall, some 64 plans, impact analyses or research reports were being completed. As one major project and then another was secured on different parts of the east coast, the fear of proliferation and sporadic development was voiced, and the need became apparent for some national land-use guidance. The Scottish Office surveyed the coast, looking for those areas which, in relation to the types of development, might best receive them, or might be most damaged. National planning guidelines were published suggesting preferred development and conservation zones on the coast which developers and local authorities should take into account when formulating their plans and decisions.

The need to coordinate activities and to communicate knowledge led to the setting up of four bodies: an Oil Development Council chaired by a Minister with representatives drawn from all walks of life; a Standing Conference of local authorities and oil companies again chaired by a Minister; a task force of civil servants; and an Environmental Forum on which representatives of all the voluntary bodies concerned with the environment considered the changing events and prospects. In addition several steering groups and working parties were set up between central and local governemnt to handle infrastructure and forward planning.

The planning machinery was heavily criticised as being inadequate to cope with the complex developments, particularly in rural areas. The local authorities were understaffed, preoccupied with reorganisation and unfamiliar with the new technology. They moved rapidly, however, and with the aid of consultants were able to prove at the time, and in retrospect it now appears to be confirmed, that when properly driven the planning machinery was adequate and indeed only four new pieces of legislation were required to control land-based development. These concerned licensing of oil refineries, designation of sea areas for construction of the later phase of concrete platforms, funding of restoration after the oil developments had gone and assistance with financial expenditure of local authorities on oil-related developments.

In spite of those achievements major criticisms continued. There were charges of the slow response of government and of local authorities; there were indeed calls for new towns at Peterhead and at Cromarty. An oil information center was demanded where anybody could find out what was going on. There were many commentators, some helpful and some over-excited. They included the Conservation Foundation of America which carried out a useful and balanced study of onshore developments in Scotland, and the Church of Scotland which issued a number of reports dealing with the reaction of communities to the oil-related development. On the other hand there was continued talk of another environmental and social rape of Scotland which, perhaps, culminated in the play which toured Scotland dealing with what was deemed to be the previous damage done to Scotland by the Cheviot sheep, then the stag and now the black, black oil.

Now that the rapid succession of planning applications has ceased, the mainland installations have been secured, and the oil and gas are flowing, some of these comments and criticisms are difficult to place in context. Indeed it might be helpful if some of the more critical commentators at the time would now look back over the comments made. A more balanced account could probably be written which might be of use elsewhere.

CONCLUSIONS

Uncertainty is one of the inevitable characteristics of oil exploration, and it leads to worry and stress. If it is recognised, something can be done. If not, speculation becomes rife, and problems are compounded. There is evidence that in the early days some companies and contractors were not able to appreciate the worry their escalating activities caused. They tended to sweep aside all local difficulties in favor of the national importance of their investments. This did not lead to trust and understanding between the communities involved and those bringing forward the new developments. On the other hand there were some outstanding examples where, in spite of the unknowns, companies, authorities and communities combined to achieve quick decisions while protecting the environment. A special effort is therefore required to communicate what is known and what is being done between the oil industry, government and the communities. New bodies may be required to establish understanding and trust between the various people concerned.

However, adaptation and extension of existing control systems is quicker than erecting entirely new ones. The British planning system with the four extra pieces of legislation was adequate to the task and the national planning guidelines provided an overview of the possible operational demand and the feasible land supply. This paved the way for contingency planning. Protecting the environment does not involve delay if there has been adequate planning and forethought. It has been said that the future can be less of a shock, i.e. less uncertain, if the various possibilities have been adequately considered. In all the oil action-areas such contingency planning took place. In addition rigorous appraisal was necessary of each project, many of which involved new technology. It was here that research and techniques associated with a phased environmental impact analysis proved valuable. The better the forward planning, the quicker the analysis, and with sufficient warning no project need be delayed waiting for the analysis of relevant issues.

The scale of proposals and the need for rapid response means that a wide range of expert advice and analysis is required, for example dealing with

    (i) need for facilities, especially platform yards;
   (ii) viability of sites for terminals and yards;
  (iii) potential social, economic and physical consequences of a project;
  (iv) infrastructure and service support program.

All this must be welded into a joint creative plan of action. Each construction scheme has its own program and activity profile, i.e. a build-up of labor activity, a plateau during which operations are running at an even level, then a curve of decline towards completion or operation. In the Scottish experience most of the programs were interrelated, and, most significantly, the rate of change, the build-up or run down, was more important than the ultimate level or plateau of activity, in terms of the effect on communities and the environment.

Finally there is, I believe, a conclusion to be drawn concerning the need to evaluate all the voices clamoring to be heard; the balance between the national priorities and community concern; the realisation that in many cases the public cannot take part or participate in the actual decision-making or technical analysis; a judgement on what the communities really feel as against the over-excitement stimulated by speculation in the news media. Scottish communities and their planners have done Britain a great service by the way in which they handled the oil-related developments. There has been a regional grasp of the challenge, and no irreversible environmental damage has been done. There has been local concern, worry and disruption, some of which still remains, but also local opportunity and enterprise. Overall, great national benefit has accrued and a great body of experience now resides in Scotland which could be of benefit to other countries.

BIBLIOGRAPHY

Aberdeen People's Press (1976). *Oil over troubled waters.* Aberdeen People's Press, Aberdeen, UK.

Architects Design Group, Lockington (1980). *Visual impact analysis for a proposed coastal reception terminal and substitute natural gas making plant near St. Fergus.* British Gas Corporation, London, UK.

Atkins Planning & Fairhurst, W. A. & Partners (1974). *Gravity stabilised oil production platforms.* Department of Energy, London, UK.

Baldwin, P. L. & Baldwin, M. F. (1975). *Onshore planning for offshore oil: lessons from Scotland.* The Conservation Foundation, Washington, DC, USA.

British Gas Corporation (1977). *Environmental impact analysis: Haddington compressor station.* BGC, London, UK.

British Gas Corporation (1978). *Environmental impact analysis: Arbroath compressor station.* BGC, London, UK.

British Gas Corporation (1980). *Environmental impact analysis of a proposed coastal reception terminal and substitute gas making plant near St Fergus.* BGC, London, UK.

Cairns, W. J. & Associates (1973). *Flotta, Orkney: oil handling terminal: an environmental assessment.* Occidental of Great Britain Inc., UK.

Cairns, W. J. & Associates (1973). *Flotta, Orkney: oil handling terminal: visual impact analysis and landscape proposals.* Occidental of Great Britain Inc., UK.

Comptroller General USA (1977). *The United Kingdom's development of its North Sea oil and gas reserves.* Washington DC, USA.

Cremer & Warner (1976). *Environmental impact of the proposed Shell (Expro) UK NGL plant at Peterhead.* Grampian Regional Council, Aberdeen, UK.

Cremer & Warner (1977). *The environmental impact of the natural gas terminal at St Fergus.* Banff and Buchan District Council, Banff, UK.

Cremer & Warner (1977). *Impact study: planning application by Shell-Esso, Mossmorran, Fife.* Fife Regional Council, Glenrothes, Fife, UK.

Cremer & Warner (1978). *Guidelines for layout and safety zones in petrochemical developments.* Highland Regional Council, Inverness, UK.

Crouch & Hogg (1973). *Gravity platforms: Firth of Clyde to Duncansby Head.* Department of Energy, London, UK.

Crouch & Hogg (1974). *Engineering analysis of alternative sites at Loch Carron.* Department of Energy, London, UK.

Crouch & Hogg (1974). *Gravity platforms: Firth of Clyde to Duncansby Head.* Department of Energy, London, UK.

Dundee, University of (1974). *An environmental assessment of Scapa Flow with special reference to oil developments.* Centre for Industrial Research and Consultancy, The University, Dundee, UK.

Economist Intelligence Unit (1975). *Buchan impact study, parts 1 and 2.* Aberdeen County Council, Aberdeen, UK.

Fraenkel, P. & Partners (1974). *Development: Loch Eriboll feasibility studies.* Sutherland County Council, Dornoch, UK.

Francis, J. & Swan, N. (1973). *Scotland in turmoil: a social and environmental assessment.* Church of Scotland Home Board, Edinburgh, UK.

Francis, J. & Swan, N. (1974). *Scotland's pipedream: a study of the growth of Peterhead.* Church of Scotland Home Board, Edinburgh, UK.

Gaskin, M. (1969). *North east Scotland: a survey of its development potential.* HMSO, Edinburgh, UK.

Gaskin, M. (1974). *The developing impact.* Royal Bank of Scotland, Edinburgh, UK.

Gostelow, T. P. & Tindale, K. (1980). *Engineering geological investigations into the siting of heavy industry on the east coast of Scotland: I, north side of the Cromarty Firth.* Institute of Geological Sciences, Edinburgh, UK.

Health and Safety Executive (1978). *Safety evaluation of the proposed St Fergus/ Mossmorran natural gas liquids and St Fergus to Boddam gas pipelines.* HSE, London, UK.

Institute of Geological Sciences (1978). *Engineering geology and site conditions of the Bothkennar area.* IGS, Edinburgh, UK.

Jack Holmes Planning Group (1968). *Moray Firth: a plan for growth.* Highlands and Islands Development Board, Inverness, UK.

Jack Holmes Planning Group (1974). *An examination of sites for gravity platform construction on the Clyde Estuary.* Department of Energy, London, UK.

Livesey & Henderson (1973). *Sullom Voe and Swarbacks Minn area: master development plan and report.* Zetland County Council, Lerwick, Shetland, UK.

MacKay, D. I. (1974). *North Sea oil and the Scottish economy.* Occasional Paper No. 1. Aberdeen University, Department of Political Economy, Aberdeen, UK.

Mackay, G. A. (1975). *Prospects for the Shetland economy.* Occasional Paper No. 4. Aberdeen University, Department of Political Economy, Aberdeen, UK.

Mackay, G. A. & Trimble, N. (1975). *Production platforms and sites.* Occasional Paper No. 3. Aberdeen University, Department of Political Economy, Aberdeen, UK.

Matthew, R., Johnson-Marshall & Partners (1974). *Nigg/Seaboard villages: development plan draft report.* Ross and Cromarty County Council, Dingwall, UK. (Unpublished.)

Mitchell, J. K. (1976). Onshore impacts of Scottish offshore oil: planning implications for the middle Atlantic states. *Journal of the American Institute of Planners,* **42**(4), 386–98.

Moira & Moira (1974). *The Lerwick local plan interim report.* Zetland County Council, Lerwick, Shetland, UK.

Morris & Steedman (1972). *Crude oil storage and ballast water treatment facilities at Dalmeny.* BP Petroleum Development Ltd, London, UK. (Unpublished.)

New Jersey, State of (1978). *The pace of oil and gas development in Scotland (1970–77): pointers for American planners.* Environmental Protection Department, New Jersey, USA.

Oil Development Council (1974). *North Sea oil and the environment.* HMSO, Edinburgh, UK.

Oil Development Council (1975). *North Sea oil and the environment: an account of the risks and action to deal with incidents.* HMSO, Edinburgh, UK.

Orkney Islands Council (1979). *Orkney structure plan.* The Council, Kirkwall, Orkney, UK.

Ross and Cromarty County Council (1972). *Towards a planning strategy for the East Ross area of the county.* The Council, Dingwall, UK.

Ross and Cromarty County Council (1974). *Impact study: planning application by Messrs Fred Olsen Ltd. at Arnish Point, Stornoway.* The Council, Dingwall, UK.

Ross and Cromarty County Council (1974). *Oil refinery at Nigg, Ross and Cromarty.* (Incorporating impact studies by consultants.) The Council, Dingwall, UK.

Scottish Development Department (1973). *North Sea oil and gas: interim coastal planning framework: a discussion paper.* SDD, Edinburgh, UK.

Scottish Development Department (1973). *North Sea oil production platform towers: construction sites: a discussion paper.* SDD, Edinburgh, UK. (Unpublished.)

Scottish Development Department (1974). *Appraisal of the impact of oil-related development.* Technical Advice Note 16. SDD, Edinburgh, UK.

Scottish Development Department (1974). *Coastal planning guidelines.* HMSO, Edinburgh, UK.

Scottish Development Department (1974). *North Sea oil: oil-related development proposals.* Circular No. 23/1974. SDD, Edinburgh, UK.

Scottish Development Department (1974). *Pipeline landfalls.* SDD, Edinburgh, UK.

Scottish Development Department (1975). *North Sea oil and gas developments in Scotland: a physical planning résumé.* SDD, Edinburgh, UK.

Scottish Development Department (1975). *North Sea oil and gas developments in Scotland: oil terminals: implications for planning.* SDD, Edinburgh, UK.

Scottish Development Department (1976). *North Sea oil and gas developments in Scotland: environmental impact analysis: Scottish experience 1973–75.* SDD, Edinburgh, UK.

Scottish Development Department (1977). *Administrative interests: planning information notes,* Series A. SDD, Edinburgh, UK.

Scottish Development Department (1977). *Land use summary sheet 6: oil, gas and petrochemicals.* SDD, Edinburgh, UK.

Scottish Development Department (1977). *Methane pipelines.* Planning Advice Note 17, SDD, Edinburgh, UK.

Scottish Development Department (1977). *National planning guidelines for petrochemical developments.* SDD, Edinburgh, UK.

Scottish Development Department (1977). *Plants and processes: planning information notes*, Series B. SDD, Edinburgh, UK.

Scottish Development Department (1978). *The social consequences of oil developments: summary of Aberdeen University research report*. SDD, Edinburgh, UK.

Scottish Development Department (1979). *Construction camps in Scotland: planning information notes*, Series C. SDD, Edinburgh, UK.

Scottish Office (1980). Offshore employment in 1979. *Scottish Economic Bulletin*, **21**, 18–25. Scottish Economic Planning Department, Edinburgh, UK.

Shetland Islands Council (1975). *Unst district plan*. The Council, Lerwick, Shetland, UK.

Shetland Islands Council (1975). *Yell district plan*. The Council, Lerwick, Shetland, UK.

Shetland Islands Council (1978). *Shetland structure plan*. 4 vols. The Council, Lerwick, Shetland, UK.

Skinner, D. N. (1977–79). *Assessment of suitability for petrochemical development*. (Series of reports on 11 sites in the east of Scotland, with emphasis on landscape aspects.). Scottish Development Department, Edinburgh, UK. (Unpublished.)

Sphere Environmental Consultants Ltd (1973). *Impact analysis: oil platform construction at Loch Broom*. Scottish Development Department, Edinburgh, UK. (Unpublished.)

Sphere Environmental Consultants Ltd (1973). *Impact analysis: oil platform construction at Loch Carron*. Scottish Development Department, Edinburgh, UK.

Sphere Environmental Consultants Ltd (1977). *Environmental impact analysis: development of the Beatrice Field*. 3 vols. Mesa UK Ltd, London, UK.

Trimble, N. (1974). *The demand for supply boat berths in Scotland 1974–80*. Occasional Paper No. 2. Aberdeen University, Department of Political Economy, Aberdeen, UK.

Trimble, N. (1975). *Demand for helicopter services*. Occasional Paper No. 5. Aberdeen University, Department of Political Economy, Aberdeen, UK.

USA Senate Committee on Commerce (1974). *North Sea oil and gas: impact of development on the coastal zone*. Washington, DC, USA.

Williams, J. T. & Conway, C. J. (1973). A study of the demand for and provision of berthing facilities in Scotland for the supply of materials in support of drilling operations in the UK sector of the North Sea. *National Ports Council Bulletin*, **4**, 8–24.

Williams-Merz (1976). *A study of gas gathering pipeline systems in the North Sea*. Department of Energy, London, UK.

Zetland County Council (1974). *Sullom Voe district plan*. The Council, Lerwick, Shetland, UK.

4

Norwegian Planning Strategy for Development of Petroleum Resources

G. GJERDE

The Royal Ministry of Petroleum and Energy, Oslo, Norway

ABSTRACT

Petroleum activities at the present level are important for Norway and should be used to make Norway a better community. The government's long-term goal is to maintain a moderate level of activity (9×10^7 tonnes oil equivalent —t.o.e.). Production in the 1980s will be $5.0–6.5 \times 10^7$ t.o.e. Development of all recent finds on the Norwegian continental shelf could produce more than 9.0×10^7 t.o.e., so it must be decided which fields are best developed. To ensure petroleum activities also benefit onshore regions Norwegian policy is to land all petroleum in Norway. Establishment of the petrochemical complex in Bamble, Telemark, and landing Statfjord gas reflect this policy. The government's concession policy, including allocation of blocks north of 62° N and in the North Sea, is briefly described stressing the importance of a stable level of exploration.

INTRODUCTION

The Norwegian energy situation differs from that in most other countries because hydro-electricity accounts for over half the total energy use. Hydro-electricity will continue to be the backbone of our national energy supply in the foreseeable future.

Petroleum production on the Norwegian shelf started in 1971. There has since been a gradual increase in production up to an estimated 5.0×10^7 t.o.e. (tonnes oil equivalent) in 1980. Oilfields under production

33

or under development will raise production to $5.0–6.5 \times 10^7$ t.o.e. during the 1980s. Since 1975 oil production has exceeded domestic needs, and in 1980 domestic oil production will amount to three times domestic consumption. Oil and gas production together are six times total national energy consumption.

The economic impact of petroleum activity is illustrated by a few statistics. This year petroleum production will contribute 14% of the GNP, it will comprise about one-third of Norway's exports and 19% of state revenue. With the production level foreseen for the 1980s, petroleum activity might comprise 15–20% of GNP[1], 35–40% of exports and 20–30% of state revenue, depending on oil prices and development of the rest of the economy.

From the beginning there has been political consensus in Norway to follow a moderate depletion policy. A production level of 9.0×10^7 t.o.e. has been regarded as a moderate production level, most recently in the government's latest White Paper on petroleum policy.[2] It is expected that the term 'a moderate production level' will be further debated in the years to come, in relation to both the national economy and the international energy situation. In the past any discussion was largely theoretical because the level of 9.0×10^7 t.o.e. was not a realistic planning figure. Development of finds on the Norwegian continental shelf until now has been decided only when development plans are presented. Now, however, several discoveries have been made where decision of development has not yet been taken.

Thus we now have the option to plan Norwegian development and production policy. It seems clear that without such planning from the authorities the Norwegian economy and society may be adversely affected.

ECONOMIC AND PRODUCTION PLANNING

Unrestricted development could lead to production levels beyond 9.0×10^7 t.o.e., and to investment in the petroleum sector which varies over the years to a such harmful extent that Norwegian industry might be unable to use revenues from petroleum activities sensibly. Such problems may be called macro-environmental effects of petroleum activities. Given this background, careful planning of development policy is imperative.

Investment in fields for which the decision to develop has been taken (Ekofisk, Frigg, Statfjord, Valhall A, Murchison, Odin, Ula and North

East Frigg) amounts to 3.5×10^{10} kroner (about US $\$7 \times 10^9$). Investment will be of the order of 8×10^9 kroner in 1980–83 and then fall drastically. Production from these fields will drop considerably from the 6.0–6.5×10^7 t.o.e. in the late 1980s or early 1990s. To maintain investment stability through the present decade and secure the necessary income in the next, new development projects must be undertaken quickly.

However, it should not be difficult to achieve the desired level of activity because several of the finds which have been made can be developed soon. First are a group of smaller projects which might be best developed relatively fast if they prove to be commercial. Reasons for this could be technical/geological, e.g. use of existing pipelines or processing capacity. While such projects will not have any decisive effect on the total level of activity on the shelf or onshore they are an important element of flexibility. They give time for careful consideration of the larger projects, e.g. development of the oilfield 34/10 (field 34, block 10, the Golden block) and the gasfields Sleipner and Heimdal. There is also a large gasfield in block 31/2 and adjacent blocks. Because only block 31/2 has been drilled, it is a rather special case and it will be some time before further development can be assessed. Landing of Statfjord gas and the work on a coordinated solution for gas production is another milestone in this policy consideration.

Today an obvious problem is that the degree of uncertainty varies according to the project. This makes it much more difficult to judge possible timing and priority of future development projects. Ideally the authorities should consider development plans for all projects simultaneously so that various important aspects can be compared:

(i) level of yearly investment and demands for manpower;
(ii) production level;
(iii) strategic market division between oil and gas production;
(iv) strategic considerations in establishing transportation infrastructure to Norway or abroad;
(v) possibilities of using existing processing or transportation capacity;
(vi) long-term policy to establish petrochemical or other industries in Norway based on supply from the continental shelf.

Because many projects are at different planning stages, this is not always possible. The importance of these projects is, however, so enormous that it is absolutely necessary for authorities to have time for detailed consideration of the most important short- and long-term implications of development decisions. This work is presently under way in the Ministry of Petroleum and Energy. The decisions taken will have effects carrying into the next century and may have important social impacts on Norway. It is therefore imperative to allow enough time for the right decision to be made.

Norwegian Landing Policy

An important aspect of Norwegian policy for development of petroleum discoveries is the rule that oil and gas from the Norwegian continental shelf should be landed in Norway. Any exception must be submitted to the Norwegian authorities for approval. The reason is that the petroleum resources should be used to create new opportunities for industry in Norway. The rule may therefore have very important environmental impacts onshore.

In spite of this general principle, no oil and gas is yet landed in Norway. The reasons are both technical and economic. From Ekofisk the gas is piped to Emden (West Germany) and the oil to Teesside (United Kingdom). The Frigg gas is also being brought ashore in the UK, and the Statfjord group was recently given permission to load oil directly into tankers. However, in permitting these exceptions to the principle of landing in Norway, the Norwegian government is in a position to impose conditions for such approval. In the case of Ekofisk, the Phillips Group had to give the Norwegian government an option on the wet gas. Wet gas components are transported with oil from Ekofisk to Teesside and, after separation and fractionation, the NGL are shipped back to Bamble in southern Norway where a petrochemical complex has been established. The complex consists of a 300 000 tons per year ethylene cracker and downstream plants for high and low density polyethylene with a capacity of 50 000 and 110 000 tons per year respectively, polypropylene with 60 000 tons per year capacity and vinyl chloride monomer production at 300 000 tons per year.

In permitting offshore loading of the oil produced at Statfjord the Norwegian government has also taken an option for the wet gas negotiations, the terms of which are to be concluded by the end of 1980. Although the Statfjord field is primarily an oilfield it also contains some gas which, as flaring is prohibited, must be produced. Plateau gas production is estimated at $4 \times 10^9 \, \text{m}^3$ per year over a period of 15 years. For the time being the gas is being reinjected into the reservoir. Reinjection can probably continue until 1985–86, when the gas will have to be taken out in order not to affect the oil production.

A further result of petroleum production on the Norwegian continental shelf is establishment of new oil refining capacity in Norway. The Mongstad refinery north of Bergen, jointly owned by Statoil and Norsk Hydro, has a capacity of 4×10^6 tons. Plans have now been submitted for an increase in capacity to 10^7 tons. This is the background for the somewhat hectic work in progress for a transport system and for further downstream plants to be ready in time for the Statfjord gas. Three alternatives for landing the gas have been discussed: the Continent, UK and Norway, of which the first two are continuously under consideration. The Norwegian solution, to pipe the Statfjord gas to western Norway, is now being evaluated. Technically this solution is feasible in spite of the Norwegian Continental Trench. Two possible landing sites (Mongstad, north of Bergen, and Kårstoe, north of Stavanger) are now the subject of local enquiries on receiving the gas and on the industrial establishments that will follow. Kårstoe requires the longer pipeline and is contingent upon the smaller Heimdal gasfield also being produced and linked to the pipeline.

In Norway we have no gas distribution system so industrial areas are under consideration. The wet gas from Statfjord is required to ensure long-term supplies for existing industry and may also provide feedstock for a new cracker in the order of 250 000 tons per year. Plans for production of up to 10^6 tons per year each of ammonia and methanol from dry gas are being evaluated.

This will still leave a considerable amount of gas. One possibility is production of electricity. More realistically, however, the gas might be re-exported as LNG or, for Kårstoe, by pipeline linked to the Ekofisk system.

A further interesting point concerning the Statfjord gas is that it may have indirect effects on development of other fields. For example, if Statfjord gas transported to Norway and the Heimdal is linked to it this will affect the timing of a possible coordinated transportation system for gas as the remaining proven reserves to be connected to such a system will be too small. If, on the other hand, Statfjord gas is not linked to other projects we still have full flexibility to develop long-term development strategy.

In the Statfjord case the Storting (parliament) will make the final decision in spring 1981 on which alternative (the Continent, UK or Norway) will be chosen. The decision will be based upon evaluation of economic advantages connected with exports compared with the industrial and other advantages of taking the gas to Norway. I have referred to plans for landing Statfjord gas in Norway in some detail because it gives a picture of possible social and environmental effects that decisions on transportation of only relatively small quantities of petroleum may have onshore. Undoubtedly Norway will face such decisions when developing other fields, e.g. north of 62° N (Lind, these proceedings).

Concession Policy

Concession policy also has wide environmental implications, and through it the government can influence the amount of exploration and thus indirectly the amount of reserves found. It can also influence the location of exploration. This is important for ensuring rational and systematic search for petroleum, taking into consideration possible transportation routes, the need for gaining experience in deeper water, etc.

The level of exploration in Norway has for several years varied between 25 and 30 wells. A parliamentary report[2] indicates that this will also be a desirable level of activity in the future. We assume for planning that reserves should be 20 times the yearly production. The reserves in blocks already allocated may be higher than this, in the range of 2·2–2·7 × 10⁹ t.o.e. One could, therefore, question the need for further exploration. However, continuous and systematic resource mapping is necessary for rational long-term planning of petroleum exploitation on the entire Norwegian shelf. Furthermore Norwegian companies might be unable to maintain and further develop their competence in exploration should this activity be discontinued. It would take years to rebuild the necessary capacity. Long-term strategy should therefore be a gradual regulation of exploration activity, not a stop-go strategy. The accumulation of known reserves could of course be regulated to a certain extent by concentrating more on high risk blocks instead of drilling the obvious 'golden prospects'.

During the period of increasing production it was natural that commercial finds very quickly led to field development. As there are several finds which could be developed now, the exploration strategy described above presupposes that the government has full freedom to decide when and how a field would be developed. Such regulation is necessary for government control of the level of activity and production. The government's right to postpone field development was first introduced for some blocks under the fourth concession round. It also applies to the three blocks allocated north of the 62° N parallel. Similar conditions must be expected in the future. Geological conditions exist for petroleum of the Norwegian shelf over about 1·4 × 10⁶ km². For many reasons, until June this year drilling only took place in the North Sea, that is about 10% of the total area of the shelf. We therefore have good knowledge of reserves in the south whereas reserves in northern areas are almost unknown to us. Long-term planning of activity and production has therefore been limited to the North Sea reserves. During the 1980s this situation will change gradually.

This summer drilling started on one block off Møre/Trøndelag and on two blocks off Troms/Finnmark, at 71°N. By 1981 it should be clearer where exploration activity should be pursued in these areas. Gradually, comprehensive plans for resource mapping over all the Norwegian continental shelf should be possible. This is a prerequisite for rational exploration planning of the entire shelf.

Blocks currently being considered by the Ministry are

  (i) blocks in 31 area;
 (ii) remaining 5th round blocks;
(iii) relinquished blocks;
 (iv) remaining 4th round blocks;
  (v) Traenabank area.

Allocation of blocks in these areas will give work for a long time to come, in particular in area 31 where a big find has been made in block 31/2, although the exact extent of the field is not known. Further drilling is therefore necessary to include this area in the long-term consideration regarding development of petroleum discoveries. An obvious problem is that the field contains very large quantities of oil in addition to gas equivalent to, perhaps, ten Frigg fields. It is not known if the oil is producible, knowledge which is absolutely necessary before taking decisions on the coordinated production of gas.

Another area worthy of mention is north of 62°N. Enough of this area must be allocated to ensure stable activity. More blocks will be allocated in the Haltenbanken and Trøms area. The opening up of the Traenabank area is also, of course, a result of this policy. Drilling there, however, will probably not take place before 1982 or 1983.

CONCLUSION

There should be no shortage of opportunity on the Norwegian continental shelf in years to come. Based on known reserves the petroleum era in Norway will last probably 50, perhaps 100, years. This creates both possibilities and problems for planning. Two aspects in particular concern impact on the environment: first, developments in the petroleum industry in Norway; and secondly, use of petroleum revenues without creating too fast a transition in Norwegian society.

For a long time it has been the policy that petroleum activities should create jobs in Norway. The setting up of Statoil, the preferential role given to the Norwegian companies Saga and Hydro and the demand that foreign oil companies should recruit Norwegians for work in Norway are all part of this policy. There is also a policy of increasing Norwegian percentage of deliveries to offshore activities. This has resulted in 36 000 people including 4000 foreigners working in the industry. The effects of petroleum activities have been relatively small. We have not seen any dramatic changes but largely a transfer from other industrial sectors, e.g. shipbuilding, into offshore activities.

The large social and environmental effect of petroleum-related activities will come when petroleum is landed in Norway and petrochemical plants, etc., are established. Norway must decide if large industrial complexes which could have serious socio-economic effects are to be established, where they are to be located, or whether petroleum revenues are to be used to establish other employment opportunities. Both are likely. There will be a move to downstream petroleum activities such as petrochemicals. The problem will then be to secure feedstock on a long-term basis. On the other hand, we will also have to use reserves from petroleum activities to develop under-developed districts.

Optimum use of revenues created by petroleum activities is a complicated subject. Revenue from petroleum activities may cause inflation, and lead to wages which may make it very difficult to export Norwegian industrial products other than oil. Wealth from petroleum activities will, unless regulated by counterpolicy, favor big industries and leave the smaller ones behind. At the same time petroleum revenue offers opportunities to transform the Norwegian society on a scale which hitherto could be only dreamed about. Whatever Norway's policy is, it will definitely be a battle ground for planners and politicians in the years to come.

REFERENCES

1. Norwegian Government National Budget, 1981.
2. Stortingsmeld No. 53, 1979–80.

5

Control Strategies: Conservation

J. McCARTHY, R. GOODIER and J. M. BOYD

Nature Conservancy Council, Edinburgh, UK

ABSTRACT

Conservation strategy in coastal lands first requires identification and delimitation of particular resources for conservation. This involves survey and inventory of conservation resources, including the abundance and distribution of species and natural habitats comprising nature conservation areas, usually listed as Sites of Special Scientific Interest, or National Nature Reserves. The results of such inventory-taking must then be evaluated and incorporated in the national statutory planning framework and other non-statutory consultation systems which enable account to be taken of conservation aims. This framework includes the system of statutory development plans and associated national planning guidelines, planning advice notes and also the consultative mechanisms between national agencies which have evolved in relation to Section 66 of the Countryside Act (Scotland) whereby those agencies are bound to have regard for the beauty of the countryside. A third element of conservation involves development of techniques for effective impact appraisal and incorporation of protective measures for the environment within the development process.

INTRODUCTION

The Nature Conservancy Council (NCC) is the official government agency responsible for wildlife conservation in Britain. The emphasis of this paper is therefore on biological conservation rather than landscape and amenity. However, there is often a close affinity between these, and much of NCC's work is carried out in close collaboration with its sister organisation, the Countryside Commission for Scotland (CCS), which deals with amenity and natural beauty. Both organisations recognise that oil developments represent the latest dramatic stage of a very long history of human influence on the ecology and landscape of Scotland.

Many oil development are located in remote areas where intensive industrial development has not previously occurred, and which have long been treasured as national resources of landscape and wildlife, relatively undisturbed by the traditional land-uses of farming and fishing. The prospects of massive oil-related developments of the 1970s sent shock waves through the conservation movement in Britain, both in the government and voluntary sectors. Much of this was related to the sheer scale and variety of these developments: onshore oil and gas pipeline landfalls; platform construction yards; oil terminals with tank farms; jetties and single-point mooring systems; gas separation and purification plants with pumping stations; oil refineries; petrochemical complexes and ethylene crackers; infrastructure of roads, housing, power supplies, etc; constant risk of oil pollution of sea and coast. All of them have great effects on the landscape and wildlife resources of coastal areas.

Control strategies for conservation in coastal lands contain three main elements:

(i) inventory and evaluation of landscape and wildlife resources;
(ii) incorporation of areas of national and regional importance for conservation into the statutory planning framework and into other non-statutory consultation systems which may also include sites of local importance;
(iii) impact appraisal and incorporation of protection measures for the environment within the development process (Fig. 1).

However, the pace of development rarely allows the sequential incorporation of these elements, and inevitably the final control strategy involves compromises to take account of many interests other than nature conservation, cometimes resulting in significant losses to wildlife. One oil pollution incident in Sullom Voe in December 1978 resulted in a known mortality of 3687 seabirds including a high proportion of localised breeding species Black Guillemot (*Cepphus grylle*),[1] while in the Cromarty Firth, current reclamation proposals will result in the loss of very substantial parts of the intertidal areas recognised as possessing international importance for their wildfowl and wading bird populations.

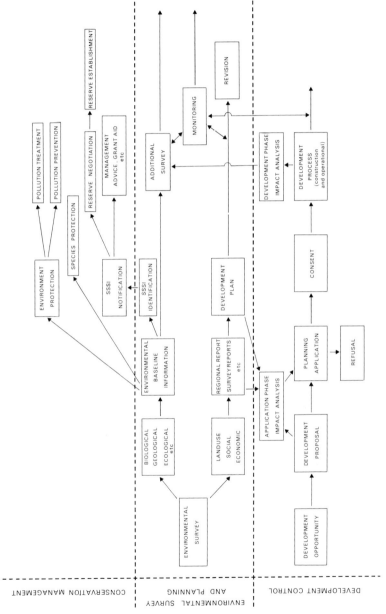

Fig. 1. Nature Conservancy Council control strategies for conservation (from ref. 6).

INVENTORY AND EVALUATION OF WILDLIFE RESOURCES

A prerequisite of any conservation policy is a clear identification of the resources which are to be conserved. For nature conservation, this means representative wildlife habitats, such as coastal sand dunes, woodlands, wetlands, etc., and the individual species of wild plants and animals. Although species conservation can frequently be achieved by a system of site protection over a range of habitats, a number of species, such as seabirds, require conservation of dispersed feeding, breeding, and roosting areas, and are sensitive to disturbance and pollution, which may have a source well outside the protected site.

The rationale for the selection of sites of nature conservation importance is set out in *A Nature Conservation Review*[2] which forms the basis for the identification of nationally important sites, some of which have become National Nature Reserves (NNRs) under Section 19 of the National Parks and Access to the Countryside Act 1949. The same Act provides for the notification of Sites of Special Scientific Interest (SSSIs) to local planning authorities, so that consultation can be carried out on any proposed developments affecting them. At present there are 53 NNRs and over 800 SSSIs in Scotland, covering the full spectrum of habitats and interests: botanical, zoological, geological and physiographical. A different set of criteria has been used by the CCS for identification of the much larger National Scenic Areas (NSAs),[3] embracing the most important landscape in Scotland.

In addition to site inventory, numerical assessments of key species are made. In the context of the present conference the best example of this is the work on seabird populations, both at their important breeding sites and also at their feeding grounds offshore as has been carrried out in Orkney and the northern North Sea. This type of census work is carried out by a wide variety of organisations such as the Seabird Group, Royal Society for the Protection of Birds (RSPB) and the Institute of Terrestrial Ecology (ITE), as well as the NCC, and involves important contributions from amateur naturalists. Another example of this type of work is the seal census work carried out by the Sea Mammals Research Unit of the Natural Environment Research Council. The process (Fig. 2) of inventory and evaluation of biological resources is complex and time-consuming, involving a very wide range of distinctive ecosystems and specialist expertise. This necessitates the use of interdisciplinary teams of marine and terrestrial scientists as was the case in the survey of the Shetland environment carried out for the NCC by ITE.[4] Often information will be adequate in one area or subject, and absent in another; yet the time

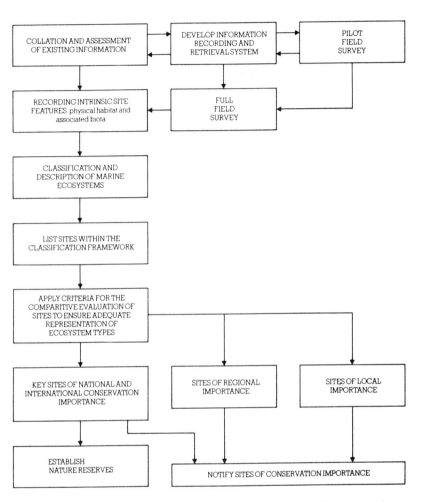

Fig. 2. Nature Conservancy Council procedure for assessment and selection of conservation areas (from ref. 17).

frame for comprehensive survey, often in inaccessible areas in adverse weather conditions, cannot keep pace with decision-taking on developments. Various measures have been adopted in an attempt to overcome this, such as mounting symposia to bring together a wide range of scientists from many institutions with specialised knowledge of the topics concerned, which has for example resulted in the publication of comprehensive statements on the natural environment of Shetland,[5] Orkney[6] and the Outer Hebrides.[7]

A more locality-oriented approach related to a potential development area was taken in the publication of a prospectus for nature conservation within the Moray Firth area[8] directed specifically at planners and developers. Apart from the identification and mapping of the main habitats of intertidal areas, salt marsh, sand dunes, coastal cliffs, freshwater lochs and others, it was necessary to obtain reliable statistics over several years (because of population fluctuations) of all the main species of wildfowl and wading birds, including their seasonal distribution and location of main food resources (Fig. 3), both intertidal plants and invertebrates, and to evaluate their importance expressed as percentages of total British and European populations of any one species. Within the Moray Firth, there is situated the Cromarty Firth, an outstanding conservation site of intertidal flats and 'soft' coast which has been for many years a scene of vigorous interaction between development and conservation employing all three stages of inventory, planning and impact analysis. By contrast, the environment of the massive oil terminal at Sullom Voe in Shetland is within a predominantly hard rocky coast with quite different marine characteristics and wildlife. In Shetland the control strategy is supported by the Shetland Oil Terminal Environmental Advisory Group (SOTEAG, which supersedes the Sullom Voe Environmental Advisory Group) which, through a system of working parties covering different aspects of the development and its impacts, produced an environmental impact assessment.[9] This concentrated on environmental aspects of the developments associated with the oil terminal, but only after decisions were made on the siting of the latter, and emphasised the need for continuous monitoring of the construction and operation of the terminal.

In breeding bird monitoring in Scotland the need was recognised to select particularly species vulnerable to oil pollution, e.g. the Black Guillemot which in Shetland makes up almost 30% of the British breeding population of the species and major concentrations of seaduck, mainly eiders *(Somataria mollissima)* and long-tailed ducks *(Clangula hyemalis)*, in the main shipping lanes. A major environmental consideration was the problem of disposal of almost $3 \times 10^6 m^3$ of peat which in places on the development site extended to a depth of 4 m. Esthetic considerations include the location of oil tanks to harmonise with the contours of surrounding terrain, sympathetic coloring, positioning and type of fencing, etc. Where time constraints completely preclude a proper survey for impact assessment there is value in arranging for experts in relevant disciplines to visit the site and give opinions on the possible impacts. While

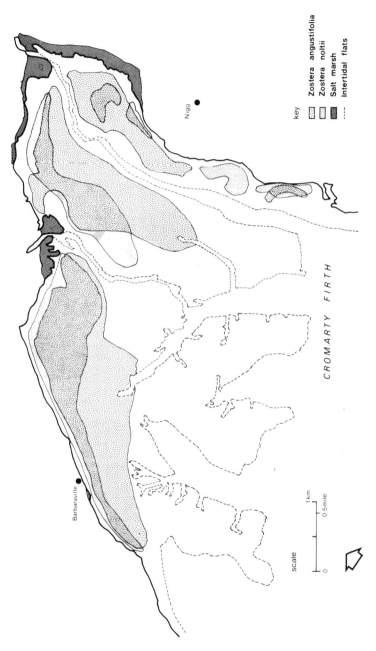

Fig. 3. Distribution of Zostera spp. and salt marsh in Nigg Bay, Scotland (from ref. 8).

not regarded as a fully satisfactory substitute for a proper survey, this approach was helpful in the case of a rapid assessment of the environmental
impact of a range of oil platform building sites which was carried out for
the Scottish Development Department (SDD) by NCC staff. The same
approach also had to be adopted in the recent assessment of the hydrological impact of the Nigg Bay reclamation proposals.

In the inventory and evaluation stages of the control strategy the following lessons have been learnt:

(i) pre-development surveys should be made, and as early as possible,
especially of populations liable to fluctuate over a period of years.
This is particularly important in separating natural from man-
induced changes;

(ii) inventory should relate to main development sites and habitats
likely to be affected by associated infrastructural works such as
roads, quarries, drainage services, etc.;

(iii) for any major developments the assembly of a multi-disciplinary
team of scientists working to a defined minimal inventory is
required, including experts in physiography and marine hydrology, particularly where intertidal reclamation is involved;

(iv) the refinement of existing information focusing on critical issues
of development impact and subsequent monitoring feasibility is
needed, while noting that readily available information has usually
been gathered in a way which makes it impossible to use it to assess
subsequent change;

(v) the determination of appropriate levels of detail required for
reliable evaluation and the focusing of attention on vulnerable
and indicator species and populations is critical; collection of unnecessary data should be avoided;

(vi) ecologists should familiarise themselves with oil development
processes and technologies in order to relate basic survey requirements to their likely impacts.

CONSERVATION IN STRATEGIC PLANNING

The main channel through which nature conservation is taken into
account in statutory planning is through the notification of SSSIs to the
local planning authorities under Section 23 of the National Parks and
Access to the Countryside Act 1949 and through establishment of some

sites as NNRs. SSSIs are only notified if they satisfy established criteria for selection, but their designation does not prohibit development; it merely ensures that NCC is consulted about proposed development (excluding agriculture and forestry) by the local planning authority, with whom the final decision on approval of such developments normally rests. Almost all sites listed in *A Nature Conservation Review*[2] (NCR) have been notified as SSSIs.

Identification of NCR key sites was done by extensive field survey of all the natural or semi-natural habitats in Britain together with equally wide consultations throughout the scientific community. These sites are the best examples of their kind in the country and are worthy of the same level of protection as existing NNRs. The total area of such key sites, however, only amounts to 6% of the land surface of Scotland; they are not by themselves sufficient to ensure the conservation of wildlife throughout the country as a whole. Indeed the value of these areas would often be adversely affected if the wildlife of their surroundings became severely impoverished. If all SSSIs are included, the land area covered by nature conservation designation in Scotland[10] is 8·5%. The aim of the notification procedure, therefore, is to ensure that through revision and refinement, as viable as possible a system of sites protected for nature conservation is created and maintained. These areas are not thereby removed from productive land-use; indeed even the NNRs and key sites are normally used for purposes additional to nature conservation.

SSSIs and NNRs represent the individual building blocks for nature conservation in the planning process, and within the period of oil development in Scotland they have become incorporated within more integrated planning frameworks. Thus in 1974 SSSIs and NNRs were taken into account in the definition of preferred conservation zones within the *North Sea Oil and Gas: Coastal Planning Guidelines* published by the SDD.[11] Inevitably there were some areas, most notably the Cromarty Firth, where such sites were preferred development zones and where it would be necessary to reach some compromise to enable the purposes of both development and nature conservation to be served. In this instance it is likely that an NNR will be virtually adjacent to a major petrochemical complex area and the NCC's efforts are now concentrated on ensuring that the siting, construction and operation of the complex are such that impacts on waterfowl and waders are reduced to the minimum through the implementation of effective planning conditions.

The NCR key sites, together with the 40 National Scenic Areas (Fig. 4) now incorporated in the Town & Country Planning (Notification of Applications) (Scotland) Order 1980 and which have a comparable status for scenic values, also have their conservation value acknowledged in the national planning guidelines which require that development proposals which are unacceptable to the NCC be referred to the Secretary of State for Scotland.[12] Both key sites and other SSSIs are also fundamental to the conservation strategy of local authorities through their development plans and are incorporated into the separate policies for industry, housing, services, etc., at both the structure and local plan level.[13] The designation of NNRs and the notification of SSSIs is part of the strategic planning process for promoting nature conservation as well as a constraint incorporated in the strategic planning process for the types of development covered by planning legislation. In Section 66 of the Countryside (Scotland) Act 1967, all government agencies in exercising their functions are enjoined to have regard to the natural beauty of the countryside, although the wording is sufficiently vague to allow for wide interpretation. There have been very few instances where it has proved possible to introduce nature conservation into development planning in the sense of relating it to the planning of those departments, agencies or companies who are actually responsible for promoting or undertaking development of major industries. This is one of the major shortfalls in nature conservation, and one of the key problems recognised by the recently published *World Conservation Strategy*.[14]

Control strategies for conservation must take account of the political and socio-economic background, particularly in areas of oil and oil-related development which have suffered severely from unemployment, poor services and depopulation. In such areas it is natural that any prospects for improvement in the local economy should be welcomed, and constraints resented. The 'ducks versus jobs' conflict arises and conservation is seen to be directed towards abstract benefits for a minority of the population, often from outside the area. A strong antipathy towards the conservation cause can develop to the point where even the most modest conservation measures are rejected. The designation of SSSIs and NSAs has been contested by some local authorities, understandably anxious that no impediments should be placed in the way of development of oil and oil-related industries. Often these reactions are based on misunderstanding of the intentions and powers of the conservation authorities concerned, but even with considerable consultation can be extremely difficult to overcome.

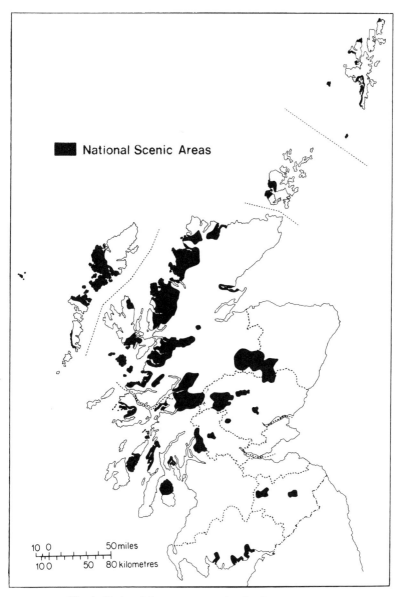

National Scenic Areas

10 0   50 miles
10 0   50   80 kilometres

Fig. 4. National Scenic Areas in Scotland (from ref. 3).

IMPACT ASSESSMENT, CONSERVATION CONSTRAINTS
AND MONITORING

A key element in any control strategy for conservation is assessment of the likely impact of development proposals on wildlife, which will enable the character and magnitude of the impacts to be determined as accurately and quickly as possible. This leads to decisions in planning whether the development should proceed at all or if it does, what constraints should be placed on the development to allow for conservation of nature. Finally, if the development goes ahead, monitoring should take place to assess if the planning predictions were correct, and to allow rapid remedial measures if any unforeseen impact occurs. Ideally all components of the process, i.e. impact assessment, evaluation and constraint imposition and monitoring, should be fully developed where a development proceeds, but this is rarely achieved in practice.

In assessing impacts, the ecology and landscape characteristics of sites are important, with a major distinction between 'soft' and 'hard' shores, and also between exposed sites on open coasts and more sheltered estuarine systems, which are often prime wildlife sites. Dependent on the type of development, impacts can affect one or more components of the ecosystem, including its geomorphology, hydrology and biology, quite apart from scenic qualities, especially in undeveloped landscapes. Although quantitative prediction of effects is usually extremely difficult some of the more obvious impacts can be identified as follows:

    (i) large-scale oil pollution by tanker wreck, leading to massive kills of seabirds and long-term damage to soft shores and their associated organisms;

   (ii) insidious small-scale continuous oil pollution either at sea or from shore bases, particularly loading terminals, affecting different components of the marine ecosystem, including benthos, plankton, shellfish, fish, seabirds;

  (iii) destruction or damage to habitats by coastal reclamation or occupation by structures and services, e.g. roads, drains, or by alteration of coast and tidal regimes locally by reclamation, coast defense works and bunds, affecting tidal flats, estuaries, spits, sand dunes, lagoons, saltings, wildfowl and wader feeding areas;

  (iv) damage to soft coasts, in particular by pipeline landfalls, affecting sand dunes, beaches, slack lakes, sea meadows, shore birds, wildfowl;

(v) disturbance to wildlife either during construction or operation of facilities; affecting breeding colonies of birds and seals in particular;

(vi) removal of material for infill and quarrying, together with spoil disposal; affecting interesting geological formations, wetlands, etc.

This is by no means a comprehensive list, but it is to the credit of the various authorities and developers concerned that most of the worst possibilities have been avoided. So far the notable exception to this is oil pollution at sea, with a significant rise in the number of incidents affecting the north coasts of Scotland in recent years, as highlighted both in the reports of the Advisory Committee on Oil Pollution of the Sea[15] and the NCC's annual reports.[16] The RSPB estimated that well over 6000 seabirds died from oil contamination around the coast of Orkney and Shetland in the first four months following the opening of the Sullom Voe terminal in Scotland. Notwithstanding the work of committees and advisory groups, much time and resources spent, and the arrangements established by central and local government for dealing with oil spills, this topic remains possibly the biggest single worry of the oil development scene. The stark fact is that in relatively remote areas where adverse weather is common, a substantial oil spill, for whatever reason, is extremely difficult to deal with, while even a relatively minor spill in a critical location or season can have catastrophic effects on important stocks of wildlife. The inescapable conclusion of most conservationists is that only close surveillance of tanker traffic and swingeing penalties for offenders can hope to contain the problem.

An effective means of impact assessment which can be linked to the statutory planning process is an essential component of any conservation control strategy, and various options for achieving this have received much attention in recent years, especially since the enactment of legislation to this end in the USA and the development of EEC proposals along similar lines (Stuffmann, these proceedings). The very complex technical and administrative problems connected with impact assessment are not examined in this paper, which simply identifies some of the key issues as they affect current practice. Because impact appraisals at present are not generally linked to any strategic planning process they are confined to the consideration of individual planning applications, and have to be accomplished within the time limits set for processing planning applications. It is not uncommon for very major developments which have undergone many months or even years of planning within the industry on technical and economic grounds to be the subject of an outline planning application which has to be processed within a matter of weeks or at most between 3 and 6 months. In such cases the impact appraisal process is seriously undermined since there is rarely sufficient information provided at outline application stage to provide a proper assessment and little or no time is provided for survey; yet once outline planning consent has been given there is a strong presumption that development will proceed regardless of any impacts which might be revealed once detailed planning applications are considered. This means that the conservation case may never be adequately presented, and that the impact appraisal is largely confined to exploring ways of ameliorating development impacts rather than playing any significant role in deciding whether the development should proceed or not or of influencing its choice of location. While this ameliorating role is certainly of value, it is important to realise that it is insufficient to prevent the continuing decline in the quality of the environment highlighted in the *World Conservation Strategy*.[14]

Where it is clear from the results of impact appraisal that a development can be accommodated without unacceptable risk to conservation interests (as was concluded by the NCC in the case of the Sullom Voe and Flotta oil terminals) then the results of impact assessment can be of great value in indicating what constraints should be placed on the development in the interests of conservation. These constraints may relate to precautions to be taken both in the construction and in the operation of the development. In the case of the current petrochemical complex proposals at Nigg Bay, NCC have recommended that certain conditions should be applied at the outline planning stage. These include an EIA giving precise information on development and likely impacts in order to specify detailed planning conditions, hydrological and ecological study to identify any necessary design modifications to the reclamation proposals which may be needed to safeguard nature conservation interests and establishment of limits for the permitted levels of atmospheric and aqueous pollutants.

A major defect in control strategies for conservation has been known for many years: the general inadequacy or absence of monitoring of the effects of development once consent has been given and the development established. In view of the great resources that have been put into arguing the effects of development at public inquiries, etc., it is surprising that so little effort is made to assess subsequently the extent to which either hopes or fears have been realised. There are, of course, some notable exceptions, such as the monitoring sponsored by the Orkney Islands Council in Scapa Flow and the monitoring of the Sullom Voe terminal under the auspices of SOTEAG. Quite apart from the function of monitoring to adjust development in the light of new information and changing circumstances relating to any particular development, the results should be of considerable value in predicting the problems of comparable developments elsewhere, and therefore of anticipating these at the outset.

Earlier in this paper the complexity and ramifications of oil developments in Scotland as they were perceived, albeit inadequately, in the early 1970s were referred to, and other contributors have referred to the NCC's unpreparedness at that time to do other than react to each new proposal in a somewhat *ad hoc* way. At the same time, this situation generated a new appreciation of the interaction of impacts on a wide variety of environmental interests within the community and focused attention on the vulnerability of Scotland's scenic and wildlife resources to large-scale modern industrial development. Perhaps it is significant that the Forum on the Environment, established by the Scottish Office in 1973 with representation from the main conservation bodies in Scotland, has subsequently broadened its remit to include all major changes in the environment, and not exclusively oil developments. This reflects a more balanced perspective on the impact of these developments arising from our experience over the last 10 years.

REFERENCES

1. Richardson, M. G. (1979). Esso Bernicia incident, Scotland. *Marine Pollution Bulletin,* **10** (4), 97.
2. Ratcliffe, D. A. R. (Ed.) (1977). *A nature conservation review.* Volumes 1 & 2. Cambridge University Press, UK.
3. Anon. (1978). *Scotland's scenic heritage.* Countryside Commission for Scotland, Edinburgh, UK.
4. Milner, C. (1978). Shetland ecology surveyed. *Geographical Magazine,* **50** (11), 730–6.
5. Goodier, R. (Ed.) (1974). *The natural environment of Shetland.* Nature Conservancy Council, Edinburgh, UK.
6. Goodier, R. (Ed.) (1975). *The natural environment of Orkney.* Nature Conservancy Council, Edinburgh, UK.
7. Boyd, J. M. (Ed.) (1979). *The natural environment of the Outer Hebrides.* Proceedings of the Royal Society of Edinburgh, UK.
8. Nature Conservancy Council (1978). *Nature conservation within the Moray Firth area.* NCC, Inverness, UK.

 9. Sullom Voe Environmental Advisory Group (1976). *Oil terminal at Sullom Voe: environmental impact assessment*. Thuleprint Ltd, Sandwick, Shetland, UK.

10. McCarthy, J. (1980). Sites of Special Scientific Interest and the Scottish landowner. *Landowning in Scotland*, No. 178.

11. Scottish Development Department (1974). *North Sea oil and gas: coastal planning guidelines*. SDD, Edinburgh, UK.

12. Scottish Development Department (1977). *Development planning and development control*. Circular 30. SDD, Edinburgh, UK.

13. Scottish Development Department (1975). *Nature conservation guidelines: regional reports advice*. Planning Advice Note No. 9. SDD, Edinburgh, UK.

14. International Union for the Conservation of Nature and Natural Resources (1980). *World conservation strategy*. IUCN, Gland, Switzerland.

15. Advisory Committee on Oil Pollution of the Sea (1979). *Annual report*. Advisory Committee on Oil Pollution of the Sea, 10 Percy Street, London W1, UK.

16. Nature Conservancy Council (1978). *Fourth annual report*. HMSO, London, UK.

17. Nature Conservancy Council (1979). *Nature conservation in the marine environment*, NCC/Natural Environment Research Council.

Discussion

F. G. Larminie. Some people might say that BP's Environmental Control Centre (ECC) is the conscience of the rich. It is a department much like any other department in BP, which is concerned with looking after environmental aspects of everything that the BP group does, from the head office in London. The group has some 70 associate companies around the world and these associates have environmental coordinators, or in the case of the larger companies environmental departments.

The ECC has developed responsibility to the individual operating companies. In provides an overview on policy and direction where problems are identified. The ECC is particularly involved in the planning of the group's activities, environmental impact assessments, contingency planning, and of course runs the group's oil spill task force. This means that the department sponsors research both in-house and externally on the development of equipment and techniques for cleaning up pollution, and development of hardware which is carried on through an associate company.

The ECC has always argued that environmental concerns are matters of science and engineering; they are not cosmetic. While public relations people may be a useful adjunct if you wish to meet the press or be advised on how best to put together a presentation, most certainly they should not be the interpretors of those matters to the public at large (see pp. 117, 120). The practitioners are the people best qualified to do this, and I reiterate my point that the ECC has consistently promoted the concept that good science and good engineering are the key elements in an effective environmental conservation policy.

Now in order to do this, of course, and not just pay lip service to the concept, it is essential that the ECC has some authority and this is done by the simple expedient of making the ECC one of the approving departments for the release of funds. The board requires ECC approval as one of the conditions for the release of capital funds for a project. This is an extremely effective method of ensuring that the ECC is consulted on all possible development projects at the planning stage.

R. A. Waller (Atkins Research and Development, Epsom, UK). In the guidelines recently published by the United Nations Environment Programme (UNEP) concerning environmental assessment of industry,[1] great emphasis is put on ensuring that decision-makers are aware of the environmental advantages and disadvantages of industrial projects. One of the problems that we had in drafting guidelines was the extent to which environmental assessment should be used to inform the community and the public at large, as opposed to informing the decision-makers.

E. M. Nicholson. I'm not entirely clear how far this aspect is separate from the normal coverage of environmental impact assessment. The problem of environmental impact assessment as it's developing here is that it is very much on a project level. Yet some environmental problems can only be discussed on a policy or strategic level. I fear that we are wasting resources by only looking at environmental impact assessment project by project.

F. G. Larminie. Mr Waller's comment really refers to two functions of impact assessment:

  (i) improvement of design, planning construction and operation. This aspect is the concern of company engineers, scientists, planners, financiers and government regulatory agencies;
  (ii) public information.

This dichotomy does lead to something of a dilemma, because there is no statutory requirement to produce an environmental impact assessment. Impact assessments are done in-house in consultation with and by

---

[1]*Guidelines for assessing industrial environmental impact and environmental criteria for the siting of industry.* United Nations Environment Programme, Industry and Environment Office, 1980 (under revision).

planning bodies, but there is no requirement to produce the results for public scrutiny.

Some companies, including BP, are starting to publish their impact assessments, something not done before more through inertia than for any other reason. Assessments are not secret and it is important not to be cut off from debate with one's peers; it would be silly to assume that we know all the answers. For example, development of the Forties field was massively debated. The internal impact assessment was used as a vehicle for explaining the project to the public and also ensured consistency in public statements by the company.

Impact assessment is clearly a useful tool but in the UK, specifically, its use should remain a matter of discretion. Mandatory use of impact assessment in public debate would approach the dangerous area of statutory imposition of environmental impact analysis. There are such major differences from project to project that it would be difficult to legislate for a standard type of impact analysis without it becoming some form of strait-jacket. That would be less productive than the present system in the UK.

A. L. Walker (Department of Management Studies, Glasgow University, UK). Mr Lyddon pointed out the enormous speed and scale of developments in the oil industry and that they and changing technology have inevitably strained the planning system. In Scotland, planning for the oil industry has been successful but certain strains were put on the system. Planners and oil companies have learnt by their experience. There are many examples, e.g. the difference between the designation of Portavadie and the planning permissions at Dunnet Bay, the new tanker route guidelines and so on. Could Mr Lyddon comment on future developments in the field of changing technology with regard to impact on the environment and also on rectifying any deficiencies in present planning procedures?

W. D. C. Lyddon. First, we should improve availability of intelligence about the future and apply it to contingency planning, which is a difficult interdisciplinary process. Some people could, for example, consider and evaluate the future technology of production platforms; they should then get together with the land-use planners and those concerned with environmental control and nature conservation so as to look intelligently at the future together. There is still some way to go to bring the right people together in order to bring their minds to bear on contingency planning.

Secondly, the new development plan system must be better used for oil-related development. Between 1970 and 1975, the peak of the development on land in Britain, the old development plan system was in force. In the new British system the strategy, i.e. the general policies for land-use in structure plans, has been separated from the particular use in local plans. Some of the lessons evident in the operation of the planning system in relation to oil-related development could be applied to the more rapid operation of the new planning system.

G. M. Dunnet (Department of Zoology, University of Aberdeen, UK). Mr McCarthy made the point that on very few occasions has there been any follow up to an environmental impact assessment (EIA). The EIA is simply a ticket to planning consent. Once planning consent has been given, it can often be forgotten. There may be a bit of cosmetic monitoring, but that is not nearly enough. At the Sullom Voe oil terminal (Shetland Islands) there have been three major changes which were never envisaged at the time of the original EIA.

First, there was no bunkering facility to be provided there, and statistics available at the time indicated the main risk was of spills of crude oil. Contingency planning was therefore related to containing, recovering and mopping up spills of crude oil. In fact the main spillages at Sullom Voe have been of fuel oil, presenting major and quite different problems, particularly in low winter temperatures.

Secondly, at the outset of environmental planning for Sullom Voe there was no word about segregated ballast tankers. Great emphasis was laid on ballast treatment facilities and so on to prevent pollution of the sea. There has since been increasing development of segregated ballast tankers. There is an altogether new environmental input which was never envisaged at the time of the original assessment. The implications of these changes must be taken seriously and dealt with to the satisfaction of all concerned.

Thirdly, since construction of the Sullom Voe terminal there has been a proposal for satellite developments. These too could not be foreseen in any detail at the time of original planning.

One initial EIA is not enough. Organisations must be established to keep watch, not just to do monitoring, so as to be aware of technological developments and to design new monitoring methods, or new activities of whatever kind, to respond to them.

J. McCarthy. Monitoring, which covers a multitude of sins, is certainly inadequate by itself. The organisational framework that will ensure continuity is equally important.

An interesting aspect of monitoring Sullom Voe is that the local authority itself is an important component. The environmental monitoring group is largely supported by the local authority. With their direct involvement there is greater likelihood of continuity of the kind that Professor Dunnet is suggesting. We must be prepared that in major developments there may be significant changes in the long-term, not just of impacts on the environment but changes in the requirements of the industry itself in that particular place. Enthusiasm tends to surge in the first 12–18 months of a project, then when it becomes apparently a more routine matter enthusiasm is difficult to find and maintain.

W. D. C. Lyddon. In EIA we have more to learn about identifying relevant fundamental issues and analysing them rapidly. Phased analysis as has been done for Flotta oil terminal (Orkney Islands) enables one to look at operational conditions. It is dangerous to generalise about EIA, e.g. by standard procedures or standard checklists. The manual drawn up by Aberdeen University[2] goes as far as one dare in this respect. The variety and diversity of projects is so great that it is much better to think it out from the beginning rather than adopt standard methods.

G. Gjerde. Norway has been little concerned with environmental impact studies, though they will undoubtedly be of greater concern in the future, and we have not yet developed methods for them. However, this will have to follow impending legislation on EIA. Norway has much to learn therefore from countries which are experienced in these matters.

E. Marshall (Shetland Islands Council, UK). It took about two years for the University of Aberdeen's research on the socio-economic impacts of oil-related development to be published, a long time for those who must plan on the basis of the research. The importance of speedy communication is emphasised by Shell's recent prediction that 60 000–80 000 jobs will be generated over the next decade by current offshore discover-

[2]Aberdeen University, Department of Geography. Project Appraisal for Development Control Research Team. *Assessment of major industrial applications: a manual.* Research Report No. 13. Department of the Environment, London. 170 pp.

ies. Has the Scottish Office considered research on what subsidiary effects can be expected from offshore discoveries and how quickly the results can be made known to those trying to plan for those effects?

W. D. C. Lyddon. There have been a number of studies on side effects, not only of oil development, but of other major industries in rural areas, e.g. by the Institute for the Study of Sparsely Populated Areas.[3] The timing of feedback is undoubtedly important. It is also a very difficult problem because economic studies labelled 'research' are often really measuring what is happening and drawing certain conclusions, and one cannot expect rapid feedback in those circumstances. Nevertheless conclusions could be drawn more quickly, although the faster conclusions are drawn the more likely they are to be wrong, so research and monitoring must continue.

Much of this work can be done in an administrative way, by logging what is happening rather than by undertaking expensive research. It should be done as part of normal joint operations between central government, local government and the oil companies.

Anon. Mr Lyddon's final remark that there has been no permanent ecological damage resulting from development in the North Sea was very firm and reassuring.

J. McCarthy. No responsible scientist can guarantee there has been no permanent ecological damage to the North Sea. Our understanding of the ecosystems involved is relatively rudimentary. When faced with oil-related development, and with having to do the sort of work that we have been, we quickly find out how little we know in some areas and how reasonably well informed we are in others. Generally the environmental interface between sea and land is most incompletely understood. With the fairly obvious exception of direct impacts on land, e.g. when a natural area has literally been removed, e.g. covered in concrete, or in the case of pollution from major oil spills, it is very difficult to identify synergistic effects, particularly in complex marine environments. Although adverse ecological effects have not so far been too bad the res-

---

[3]Institute for the Study of Sparsely Populated Areas, University of Aberdeen, an interdisciplinary research unit concerned with social and economic problems of marginal regions in all countries around the North Atlantic.

ponsible ecologist and scientist must reserve judgement on permanent effects until we have a great deal more experience. Ten years is a very short time indeed.

K. Stenstadvold. An interesting aspect of the careful and seemingly systematic planning in Scotland is that the coastline of the whole country has been considered. This has been attempted too in Norway but the search for suitable sites for petrochemical development has special problems. When government agencies or oil companies look at potential sites in a purely speculative way local authorities tend to interpret it as a promise to develop the land. There is relatively little concern about the natural environment while potential jobs are looked on as much more important (see Gjerde, these proceedings). In Norway this means that as a result of these speculative surveys mayors from local communities claim to have been promised some development, making such strategic planning difficult. Do local authorities in Scotland interpret speculative surveys as a short-list of preferred sites?

W. D. C. Lyddon. The terms 'preferred conservation zone' and 'preferred development zone' are used very deliberately. They do not represent government command or government policy, but a view at the national level of the coast of Scotland. On such a wide scale, of course, detailed resolution is not possible and can only attempt to equate national demand with national supply.

Preferred zones are a form of national guidance, advising that industrial development in a preferred conservation zone is likely to encounter environmental issues, problems and objections. Conversely, parts of the coast already urbanised and not having outstanding ecological and scenic value are labelled, in an advisory way, as preferred development zones.

The flexibility of this designation allows adjustment if the division between development and conservation zones is wrong. The designation is thus part of the planning dialogue.

J. Balfour. Mr Gjerde reminded us of the importance of clear objectives for dealing with development and using oil revenue in the interest of the people concerned, and to remember the importance of planning without haste. We have been reminded also of the value of identification of biological resources, whether on land or in sea. It may not necessarily be the amount of land that is used in development or the area of sea, but it may be where these are and what kind they are.

The importance of recording and remembering what happens so that we can do things better in the future has been emphasised. There may be a need to put greater resources into record keeping and the research that may need to go with it than has been the case in the past.

Section II

OIL, STRATEGY AND DEVELOPMENT

Chairmen

Papers 6 and 7
Professor Sir Kenneth Alexander, FRSE, Chairman,
Highlands and Islands Development Board, UK

Paper 8
H. Hunter Gordon Esq., Chief Executive, Ferranti
Offshore Systems Ltd, Edinburgh UK

Papers 9 and 10
Professor T. D. Patten, FRSE, Acting Principal,
Heriot-Watt University, UK

6

Mossmorran: Planning Considerations†

J. F. TAYLOR

Shell UK Exploration and Production, London, UK

ABSTRACT

*The Mossmorran natural gas liquids (NGL) fractionation plant and Brae-
foot Bay marine terminal will be the last parts to be completed of an extensive system to make gas and gas liquids from the Brent field in the North
Sea available for commercial use. The Mossmorran site is in an area
designated for industrial development and is in a part of Scotland with infrastructure capable of supporting major industrial expansion. Braefoot Bay
is an unusually fine site for a marine terminal, providing an excellent deepwater location shielded from other marine traffic in the Firth of Forth. Both
sites have the additional attribute of being topographically shielded from
the surrounding area.*

*Stringent standards have been set by local authorities and by the company
to virtually eliminate the environmental impact on nearby communities.
Safety factors have been paramount in design of the plant and the marine
terminal considerations.*

*Acceptance by the entire community is another kind of planning considera-
tion. In the construction phase, links must be built with local business interests
and with those people in the immediate area of the sites who may be affected
by construction activities. 'Planning' must continue when these new facilities are in operation. For an industrial development to be successful in the
long-term, planning considerations must not only establish the base on which*

† This paper was presented by Mr J. Carder, Shell UK Exploration and Production, Shell-Mex House, Strand, London WC2R ODX, UK.

*to build but must represent an ongoing activity so that the new development
and local communities all prosper together.*

INTRODUCTION

The Brent field was discovered in 1971 by Shell Expro (operator in the
North Sea on behalf of Shell and Esso), and was quickly recognised as the
UK's biggest single oil and gas field. It has the highest gas:oil ratio of
any UK field currently under development and is estimated to contain
recoverable reserves of 2×10^9 barrels of oil and NGL and $3.5 \times 10^{12} \text{ft}^3$
of natural gas, sufficient gas to provide some 15% of the UK annual
demand.

The Brent field contains associated gas and oil, and substantial quantities of gas liquids (hydrocarbons whose gaseous or liquid state can readily
be reversed by small changes in temperature or pressure). In developing
the Brent field significant capital investment (over £750 million of the
total £3400 million) will be incurred to collect and process the gas and
gas liquids.

Oil from the Brent field is already being piped to Sullom Voe in Shetland. By the end of next year the first gas and gas liquids are expected to
move down the 278 mile submarine pipeline from Brent to St Fergus
where methane will be separated and sold to the British Gas Corporation
(Fig. 1). When construction in Fife is complete, the remaining mixture
of gas liquids will be piped from St Fergus to Mossmorran for fractionation and processing into marketable products to be shipped from the
Braefoot Bay terminal. The Mossmorran fractionation plant and Braefoot
Bay shipping terminal in Fife are thus the final links in a chain of investments to make the gas and gas liquids available for commercial use.

SITING REQUIREMENTS

Clearly the siting of such important links in such an extensive chain is not
lightly considered. The large volume of gas liquids necessitates provision
of export facilities, and this means a shipping terminal. There are, therefore, a number of interdependent factors to be considered in the search
for a suitable location. In particular it should be in an area designated
for industrial development and have the infrastructure necessary for
subsequent downstream development. It will need the full support of

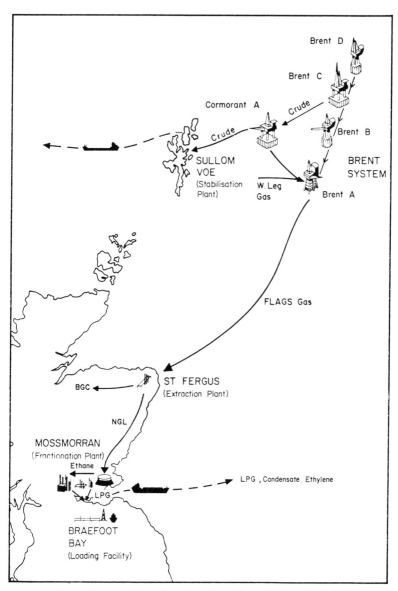

Fig. 1. Far north liquids and associated gas system—FLAGS.

local authorities. It must have access to a marine terminal which itself must provide an unrestricted all weather deepwater harbor. The jetty should be remote from passing traffic and be protected from the full force of wind and wave. Clearly, it is no good finding an ideal terminal location if there is no suitable plant site nearby, or vice versa. Equally clearly, it will be almost impossible to find a location which will completely satisfy all the criteria. Safety (see below) and environmental impact must also be considered.

The Firth of Forth is a sheltered deepwater port whose natural features are enhanced by the existence of a strong and well established port authority. Although there is substantial marine traffic already using the Firth, none passes through Mortimer's Deep, the deepwater channel separated from the main estuary by Inchcolm Island (Fig. 2). Braefoot Bay itself provides relatively close access to this deepwater channel and Inchcolm Island provides a shield from other marine traffic. The land site adjacent to Braefoot Bay has the further attribute of being most effectively screened topographically.

Whilst the 4–5 mile coastal strip inland from Braefoot includes some of the most picturesque countryside in Fife, the Mossmorran site is completely concealed from the south by the Pilkam Hills and, seen from the north, is adjacent to the extensive opencast coal workings of the South

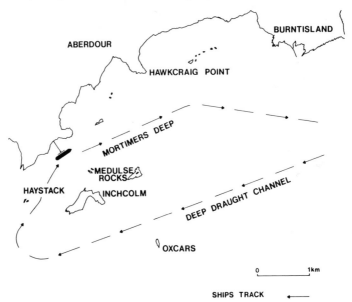

Fig. 2. Deep draught channel access to the Braefoot Bay terminal, Firth of Forth.

Fife Coal Measures. This area was identified some years ago by the local authorities as suitable for industrial development, as part of the effort to counter declining employment on the coal workings. Because of its long industrial history, Fife has a developed infrastructure of power, transport and other services as well as access to supplies of labor for construction and operation of the fractionation plant, the ethane cracker and any future downstream industries.

Locating any plant of the size and complexity that Shell is building in Fife will inevitably have some effects on the local environment. In his 1979 grant of consent for building these plants, the Secretary of State for Scotland imposed almost 50 planning conditions of which seven relate to safety matters, 15 cover air and water pollution controls, nine are concerned with visual impacts, and the remainder represent general controls on the development. The Secretary of State also invited the Company to make special provisions, outside the planning legislation, for activities and property outside the development sites by

(i) improvements in visual screening through offsite landscaping;
(ii) arrangements with the local authority for housing the nearby Gray Park community;
(iii) minimum interference with the popular annual Aberdour Regatta;
(iv) consideration of special problems posed to a smaller number of owners of adjacent properties;
(v) the formation of a Liaison Committee (see below).

Furthermore Shell has entered into a legally binding agreement with the local authorities under Section 50 of the Town and Country Planning (Scotland) Act concerning several of the above. The result of these controls and agreements will minimise impacts due to pollution and visual intrusion on neighboring communities. For example, although there have been minor problems during site preparation, water in the Dronachy Burn that drains the site will eventually be cleaner when it leaves the site than when it enters. Also, landscape architects have taken considerable pains to enchance the natural screening by construction of mounds and by tree planting schemes.

SAFETY

Although superficially less obvious, but of fundamental importance to neighboring communities, safety has been a paramount consideration in design of both fractionation plant and marine terminal. Most facilities being built at Mossmorran and Braefoot Bay will exceed national and international safety code requirements since they must also meet all the requirements of Shell and Esso, which are often more demanding. The whole design is based on a fail-safe philosophy providing assurance of maximum safety whatever the malfunction, for example

  (i)   double-walled storage tanks for low temperature products at Mossmorran, assuring product containment even in the unlikely event of failure of the inner tank;
 (ii)   tanks that are partially embedded in the hillside and surrounded by a rock and earth embankment;
(iii)   double security fence all around the plant;
 (iv)   comprehensive fire detection and fire-fighting systems including fire-water mains, two deluge systems plus foam, dry powder and inert gas systems; a depressurising system that enables controlled pressure reduction of the process plant in an emergency such as a fire.

No finished products from the fractionation plant are stored at Braefoot Bay, but the terminal has its own safety features to assure safe and expeditious loading of the special gas ships which will carry low temperature export products. An automatic device will shut down loading pumps and disconnect loading arms, without loss of product, in an emergency. Such disconnection could take place as a result of excessive movement of the ship being loaded, or may be initiated by ship's crew or by shore personnel. On the jetty a system of surge drums has been arranged to remove all liquid and vapor products from the jetty head prior to departure of the ship, leaving the loading lines and arms full of nitrogen, ready for berthing the next ship. Only ships constructed and maintained to the highest international standards will be permitted to use the jetty. Every ship coming into the Firth of Forth to load at Braefoot Bay will have its record checked by the Forth Ports Authority and will not be allowed in unless the harbor master is satisfied with its record.

The most important contribution to safety, as always, will be made by the people who work at the installations. They will have special training in use of the facilities and will be made constantly aware of their responsibilities. Great emphasis will be laid on 'good housekeeping'. The types and amounts of safety equipment and the safety procedures themselves will be subject to regular review. Safety is integral to the whole operation from the way the facilities are designed and constructed to the way they are maintained.

PUBLIC RELATIONS

A further planning consideration involves the Company's relationship with surrounding communities. Local concern over safety is understandable, and the task of reassuring people about how plants will be constructed and will eventually operate is of prime importance. No business can operate successfully without acceptance by the majority of local people, and to achieve this for Mossmorran and Braefoot Bay requires patience and a determined and continuing effort. Objections to the development by individuals may not be entirely overcome; in so many cases they are emotional and difficult to counter, but fears based on misunderstandings can and must be alleviated. A program of communication with local communities to provide general information about the project and to deal factually with matters of safety is essential. This must be a genuine attempt to satisfy concerns over safety, by the presence of qualified and experienced personnel at all meetings.

During the construction phase at Mossmorran and Braefoot Bay close links are being built with local businesses to ensure that as much work as possible is done by local labor. In conjunction with Fife Region local meetings were set up at the outset at which the managing contractor Ralph M. Parsons met and talked with nearly 800 industrialists, to explain the project and the sort of materials and services that would be required. Tendering and contract procedures were dealt with and the names of contacts given for obtaining further information. Full and fair opportunity will be given to Fife firms in the hope that they will be successful in their bids for business which must of course be competitive. The more successful they are, the more local employment there will be. This will generate more spending power locally and thus add to the general economic well-being by recycling newly created wealth. Through close liaison with Fife Region the job centre at Cowdenbeath has been upgraded to coordinate all requests for work on the project that are received at other job centres in Fife. Subcontractors are required to consider local labor for work before going out of the Region to find it, and many contractors use the Cowdenbeath job centre to carry out interviews with potential employees. The behaviour of contractors must be exemplary at all times; they in effect are representing Shell and by their actions and the actions of their subcontractors are one of the Company's interfaces with the public. The developers will be judged by the contractor's performance, and the Company therefore has a need to ensure that contractors are fully aware of the business ethics and behaviour that govern business conduct.

To deal with any problems that may arise a proper complaints procedure is being established, to ensure that all matters are speedily and acceptably dealt with as they occur. Nothing rankles more than to have one's legitimate complaint seemingly ignored. To minimise those occasions when complaints are made, the public will be as fully informed as possible about the various stages of construction, particularly those that will directly affect the local community. Experience shows that warning of an event, not just what it is but why it is happening and what place it has in the total project, can make even an unpalatable occurrence acceptable if the need for it is properly understood. It is this understanding on a personal level that establishes a sense of trust in the Company and its activities.

The process of keeping the public and the news media fully informed is a continuing one during the construction phase. Literature explaining all about the project needs to be readily available, and informal meetings should be held at intervals with representatives of local councils. As soon as consent for the project was received from the Secretary of State for Scotland, meetings were held with the full councils of the Region and the Districts, and subsequently with community councils, to explain proposals in detail and to answer questions.

As construction nears completion and the operating phase approaches, explanations of the plant process, how it works, what it does and what it produces must be made available to all who live nearby. An effective way of doing this is through site visits, when visual presentations can be made and operating personnel are on hand to answer questions. The emphasis is the same throughout the construction, commissioning and operational phases: to provide information to achieve understanding so that the whole operation can be seen as it is, a useful and considerate neighbor, and to enable it to function with public support and acceptance.

Once the plant has been commissioned and is in operation, the process of communication becomes a regular feature, a 'planning consideration' which has not only established the base on which to build but which also represents an ongoing activity so that the new development and local communities all prosper together.

7

The Search for Oil and Gas on the Continental Shelf North of 62°N: Present and Future†

T. LIND

The Royal Ministry of Petroleum and Energy, Oslo, Norway

ABSTRACT

Offshore drilling for petroleum above 62°N started this year, after a long period in which political hurdles contributed to continuous postponement. Reasons for opening this very large and potentially productive area for exploration are presented, as well as the conflicts involved. In 1980 three wild-cat wells were spudded in two offshore areas. The results are presented, and the organisation of exploration activities is described. Onshore impacts, especially those connected with supply bases, and the results of efforts to initiate industrial and research projects between oil companies and local firms and institutions are examined. In the light of experience in 1980, plans for further drilling offshore north and mid-Norway are discussed. Important issues include the expected number of wells, possible extension of the drilling season, and whether or not new licenses should be allocated and new offshore areas opened for exploration. Finally on the basis of recently published comprehensive research material, the possible effects of future petroleum discovery above 62°N are considered. The paper focuses on the onshore impact of landing and processing oil and gas in north Norway.

† The views expressed in this paper are not necessarily those of the Ministry.

INTRODUCTION

Drilling for oil and gas to map the possible resources of the Norwegian continental shelf above the 62nd parallel was proposed by the Norwegian government early in the 1970s, only a few years after the first commercial discovery was made in the Norwegian sector of the North Sea. In 1980 petroleum exploration started. Drilling of the first three wildcat wells, of which two are already finished, started in May in two separate areas. The northernmost area is at 71°N, about the same latitude as oilfields in northern Alaska and Canada. The other is at about 65°N.

Until summer 1980, Norwegian petroleum exploration was confined to the North Sea. So far 260 wells have been spudded in, of which 195 are wildcats and the remaining 65 are delineation wells. Proven recoverable reserves and other estimates are given by Gjerde (these proceedings). Norway's continental shelf is larger than the North Sea, and is, according to international law, about $2 \times 10^6 \, km^2$, of which $1.4 \times 10^6 \, km^2$ are suitable for petroleum exploration. Norway's continental shelf north of the 62nd parallel is about ten times the size of Norway's sector of the North Sea.

This paper concentrates on north Norway, i.e. the counties Finnmark, Troms and North Nordland, although much of the discussion is also applicable to mid-Norway. The three north Norway counties cover about 120 000 km² with a population of nearly 470 000 (four people per square kilometre average). Most people are in small towns and villages on the coast.

Drilling was undertaken for three main reasons. The most fundamental was to discover as much as possible about Norway's natural resources. In addition, since the war, north Norway has experienced economic decline and emigration. Petroleum exploration and production is a means of furthering the economic development of the region.

The third reason for drilling in the northern waters was simply to find enough petroleum to secure a comfortable planning horizon for Norway. Before the fourth round of licensing (i.e. before 1978) there was concern that production of oil and gas would fall sharply in the 1990s, as no substantial discovery had been made for some years. The fourth round, however, will bring annual production to 9.0×10^7 tons oil equivalent (t.o.e.) or more in the 1990s if all fields are developed. Although this figure is regarded by the government as a 'moderate level of production' it can, in practise, be regarded as a production ceiling. This argument for exploration north of the 62nd parallel is not, therefore, convincing today.

PUBLIC CONCERNS ABOUT OIL DEVELOPMENT

In spite of apparently good reasons, at least when they were presented, it has been difficult through the 1970s for the various governments to open the shelf in the north for drilling. Postponement has been because of anxieties about environmental protection, particularly the safety of operations, preparedness for oil spills and opposition from the fishing industry. Postponement may have, in many ways, been justified—the time was perhaps not yet ripe for moving into the offshore area before 1980. Three sets of arguments finally made the step possible:

(i) a steady increase in the development of offshore safety measures throughout the 1970s, and an acceptance by the Storting (parliament) and the general public that 100% safety is unattainable. Of course in some quarters the safety level is still unacceptable;

(ii) the Bravo and Ixtoc-I blow-outs and the wrecking of several big tankers indicated that oil spills do not have long-term disastrous effects on fish stocks and marine life in general. This has largely been confirmed by research. The change in understanding of the effects of oil pollution came at the same time it became publicly known that the oil booms Norway depends on as the main strategy to fight offshore oil spills did not meet the standards set by the authorities. This realisation was, of course, fortunate for the government, and the point was not forgotten by the Industry Committee when it recommended the Storting to vote in favor of oil exploration in the north;

(iii) the third obstacle was the vocal and politically strong lobby of fishermen. Their objections were very much connected with the above, but the fishermen also had special interests. Against the background of what had happened in the North Sea in the pioneer days of oil exploration, the fishermen had some reason to be worried. Fishing is the backbone of the economy in most regions of north Norway, and scraping of the sea floor by the oil industry on the scale done in the North Sea could not be tolerated.

Realising this, the government urged clean up of debris from drilling sites and supply boat sailing routes, introduced a new law on pollution of the sea floor and, perhaps most important, recognised that fishermen had legitimate interests in offshore areas. As a result fishermen's organisations have been involved in planning oil-related activity. In the north, they have been involved in decisions on the choice of areas for drilling, on the level of activity, on restricting drilling to summer, on designation of supply boat routes and on other matters. The Ministry of Petroleum and Energy now holds regular consultations with fishermen's organisations on all matters of mutual interest.

While such involvement has made the oil industry more acceptable to the fishermen, the lack of fish in the sea, partly the result of overfishing by the fishermen themselves, has emphasised the need for additional employment opportunities in north Norway. Both central and local government look to the oil industry to help achieve it, at least in the long run.

To create new jobs early in the oil era, and to obtain industrial and technological help from abroad, the oil companies have been asked to enter into joint research and industry projects with local firms. This has been part of concession policy in the two last rounds. Although the companies have signed agreements which should give some 200 new jobs in north Norway, few jobs have actually materialised. It seems very difficult to stimulate rapid economic growth in this way, and in the short-term other measures must be found. The government is now launching an 'Action Plan for north Norway' in which the Storting is asked to spend several hundred million kroner on a crash program to enable north Norway to ride the coming crisis. Whether or not the oil activities will give north Norway an economic platform depends upon discoveries being made. The exploration phase does not create enough jobs for locals, either offshore or in supply bases which employ some 10–20 people, nor does it induce any multiplier effect.

If the current search for oil and gas in the north is regarded in this context, it is quite easy to understand why the drilling of three first wells has attracted so much public interest. The aim of the wells has been primarily to test various types of geological province in order to establish the possibilities of hydrocarbon formations being present. It was therefore of great importance when the wells drilled this summer clearly demonstrated that the right conditions were present in the Troms I area, where Norsk Hydro and Statoil have just finished their wells. It is established that the three prerequisites for the formation and accumulation of hydrocarbons are present, i.e. (a) the right geological structure, (b) the source rock and (c) the reservoir rock. The question is, therefore, no longer if a petroleum discovery will be made, but when. The task for the years to come is really one of finding the right combination of the three prerequisites. It is no secret that Statoil was close to success with their well this summer. The company has had shows of hydrocarbons which their geologists classify as 'very interesting', but they are reluctant to describe it as a find. Since the drilling season in the north is now over, testing the well must be postponed until next year. In 1981 more wells will be spudded in other types of province in the same blocks. One can only hope that the results will match the high expectations.

POSSIBLE ENVIRONMENTAL EFFECTS OF DEVELOPMENT

What then are the scenarios of a commercial discovery off the coast of north Norway? What environmental, economic and social impact will the development of a field have upon the affected regions?

These questions were included in the mandate of a Royal Commission appointed by the government in 1976 to investigate the possibility and consequences of a petroleum discovery above the 62nd parallel. A further task of the Commission was to initiate and coordinate research into the possible effects of landing, processing and transporting petroleum. The Commission's report is a comprehensive document covering a wide variety of subjects. Of special interest here is a discussion of alternative uses of oil and gas landed in north Norway. The report mentions four main alternatives: export, petrochemical industry, non-petrochemical industry and power production.

To be commercial a discovery off the coast of north Norway must be very large. Since the market for petroleum products in north Norway is negligible, it would not make economic sense to process all petroleum locally. Consequently, most of the petroleum will have to be exported. That is easy enough with oil; it may even be possible to export the crude directly by offshore loading without it ever reaching the shores of north Norway.

With gas, however, preliminary studies conclude that it is technically feasible to lay sub-sea pipelines from the current exploration areas to the coast. Unlike the UK or continental countries, we do not have a national gas grid. Other ways of disposing of the gas must therefore be found. In principle there are three options:

(i) a new 2300 km pipeline across the Scandinavian peninsula to the continental gas grid;
(ii) ship gas as LNG to the overseas markets;
(iii) produce massive quantities of bulk petrochemicals for export. This is an alternative which will be considered in any event if condensate is landed.

The government is currently looking into the feasibility of the alternatives or of combinations of them. For instance, a group of Nordic civil servants and experts are now studying how a Scandinavian gas line can further the development of north Scandinavia and improve coordination between the countries. If a pipeline is considered feasible it will be necessary to spend a great deal of time and money on assessment of the environmental and social consequences of such a gigantic enterprise.

Detailed discussion of the other alternatives is outside the scope of this paper. In general, however, petrochemicals are generally regarded by the government as something Norway should produce. The difficulty is that they are normally produced in complexes far too large to fit easily into the sparsely populated regions of north Norway. It would be difficult for a developer to promote a petrochemical complex, with its demand on land, capital and highly skilled labor force, as a solution to the problems of low incomes, high unemployment and emigration in a fishing community. Theoretical attempts to split a complex into smaller separately located units are not encouraging. One study indicated additional costs of about £50 000 per job per year for a decentralised plant. It may be relevant to question if money might not be better spent directly on regional development.

POSSIBLE ECONOMIC GROWTH FROM OIL-RELATED ACTIVITY

North Norway is presently in a pre-discovery phase, of which an important characteristic is uncertainty. No one knows when or where a discovery will be made, how large it will be or of what it will consist. The first aim is, therefore, to reduce this uncertainty, which can only be done by further exploration, i.e. more wells and exploration in more blocks. This is also in the interest of national development. The oil and gas potential of a number of blocks should be thoroughly mapped before planning comprehensive development and transportation systems. Such investigation would also make a stable level of production possible, which is important at least from the transportation and onshore processing point of view.

Since it will often be in the interest of the oil companies to get the fiields in production as fast as possible, there may be a conflict of interest between the oil companies and the government. The government has therefore taken powers to postpone development of fields found north of the 62nd parallel indefinitely, without having to state reasons.

However, when the resource base is established and the transportation and processing alternatives are agreed upon, there are many good reasons for going ahead with development but within the government's production ceiling of 9·0 × 10^7 t.o.e. per year. Currently a number of very large fields in the North Sea are ready for development. If all are developed, there will be no room within the production ceiling for north Norwegian fields before the year 2000. It does not take much imagination to foresee a political debate on whether or not north Norwegian fields should be given priority and, if so, which North Sea fields should have development postponed.

In any event, the development of fields off north Norway, and indeed also the discovery of them, is a few years ahead. The time available is a resource which should be spent wisely, because it gives the opportunity to analyse possible consequences of oil-related development; it enables local industry and labor to prepare for opportunities to come; it gives central and local government a chance to draw up plans and discuss strategies for the development of the use of resources.

Through a number of research projects and internal team work the industry has been preparing comprehensive documentation of onshore and offshore impacts of all phases of oil activities in north Norway. The studies are undertaken to answer such question as

(i) what are the possible effects of a large oil spill upon marine life, especially fish, off north Norway?

(ii) is there any conflict between oil industry and fishing, particularly with regard to recruitment of labor?

(iii) in what ways can agriculture be affected by the oil activities?

(iv) is it possible, in advance and on a theoretical basis, to identify the types of local community best suited to various kinds of oil-related developments?

(v) can the special working hours in the offshore oil industry, with one week in four at home, make it possible for the oil employee to keep a part-time job in, for instance, fishing or farming?

Environmental impact assessment is not yet compulsory under Norwegian law. Nevertheless, the importance of research and impact analysis in the pre-discovery phase can hardly be stressed too much. At their best, such projects function as an early warning system, allowing authorities to react accordingly.

As a complement to research projects it is advantageous to monitor effects of oil activities as they occur. The Norwegian government is sponsoring programs which give a continual flow of data on the ecological and biological status of marine life at drilling sites, and on that part of the coast which might be affected by a large oil spill. Pre-drilling data was collected in the same areas. A further monitoring program registers use of local labor, goods and services by the oil industry. The latter is critical in north Norway where the manufacturing industry is small and oriented towards processing of local resources, e.g. fish and minerals, or servicing of local markets. North Norway is poorly suited to supply the oil industry with goods and services. Upgrading and a shift in orientation is essential in the light of the new opportunities.

A government commission recently concluded that a whole range of incentives were necessary to prepare the north Norwegian manufacturing industry for the oil era. Suggestions include

  (i) increased transportation subsidies;
  (ii) financial support for job training;
  (iii) positions for oil consultants at the county administration;
  (iv) cooperation and joint ventures between local firms.

The Commission stressed the importance of north Norwegian firms providing goods and services for oil-related activities in the North Sea, thereby building up competence and capacity to meet future development off north Norway. Early in 1981 the Norwegian government will propose ways to increase north Norwegian participation in the oil activity to the Storting.

When new plant is being built it is commonly regarded as an advantage to try to achieve high local employment, at least in the operational phase. The oil industry primarily needs people with higher education and special skills. There are few such people in the north. Since the training of labor often takes several years, even when proper facilities are available, the sooner the process is started the better. The government department responsible for education is now planning for training of people for oil jobs, but a large proportion of north Norwegians in the oil industry cannot be expected for a number of years.

The pre-discovery phase gives central and local government the opportunity to plan for development and disposal of future petroleum discoveries. In north Norway, petroleum is an entirely new resource, as it was in Norway only 10–15 years ago, and few people know much about it. While it is sensible for any region facing oil-related activities to build competence on the subject in advance, it may now be too late as one developer is already filing a planning application with the local planning department. One way of building competence is to model the possible environmental, economic and social effects of a simulated oil and gas discovery, which are then discussed in the various departments involved at the central, regional and local level. Such model studies have already been undertaken by central government, and some regional authorities are now following suit.

CONCLUSION

If the time available is used actively and sensibly by the local industry, the labor force and the authorities, north Norway should in many respects be in a better position to accept and develop with an oil and gas industry than was Norway when petroleum was first discovered in the North Sea.

8

Forties Field Development: The Environmental Aspect

M. M. LINNING

32A Fountainhill Road, Aberdeen, UK

and

F. G. LARMINIE

*Environmental Control Centre, British Petroleum Company Limited,
London, UK*

ABSTRACT

*The Forties field, developed by British Petroleum, was the first major oilfield
to be discovered in the North Sea (1970). The strategy adopted by BP during
and after development of the production system of that field, comprising
offshore platforms, submarine and land pipelands, storage and export
facilities, included full environmental impact assessment. The safety of
people and their environment was a foremost concern in implementation of
this major offshore development, in which the underlying theme was one of
reasonable compromise and rational debate.*

INTRODUCTION

The Forties field, discovered in October 1970, was the first major oilfield
found on the UK continental shelf. Appraisal drilling in 1971 confirmed
the approximate reservoir dimensions and established that it contained
recoverable reserves of 1.8×10^9 barrels of 37° API oil, which made it a
major field by international standards. Coincident with field delineation
work feasibility studies were in hand to determine how this oilfield,

located in the northern North Sea, 110 miles east-north-east of Aberdeen in 420 ft of water, should be developed.

Engineering studies considered development of the field by fixed platforms, feeding to an offshore loading scheme, but that was discarded in favor of a submarine pipeline to shore, a solution which was demonstrably economic given the size of the reservoir. Also, siting of the onshore terminal was analysed and alternative sites evaluated, but considerations of probable cargo destinations and the advantage of integration with an existing refinery gave clear preference for the Fifth of Forth (and the long-established Grangemouth refinery) whose estuary provides one of the best protected anchorages on the east side of Scotland.

After evaluation the preferred method of development was fixed platforms spaced to provide reservoir cover by deviated drilling. The platforms were to be connected to a 32 inch submarine pipeline making the shortest landfall 105 miles from the field at Cruden Bay, thence by a 36 inch landline down the east coast of Scotland to Grangemouth, where the oil stabilisation and gas treatment facilities would be sited with provision for delivering the stabilised crude to a loading terminal on the south bank of the Firth of Forth.

The decision to develop was made by the Board of British Petroleum in December 1971, and the first oil flowed in September 1975. The field currently produces about 500 000 barrels per day and to date has produced more oil than any other field in the North Sea.

The environmental conservation and protection policy employed throughout development of the Forties field (1971–75) had the following objectives:

(i) the design, engineering and construction of the facilities (e.g. field production system, pipelines, storage, treatment and loading system) had to have integrity under all foreseeable operating conditions;

(ii) the location, route and configuration of the facilities should avoid disturbance of the existing environment, land and communities as much as is reasonably possible;

(iii) the consequences of a systems failure due to error or due to conditions that could not have been foreseen had to be considered and practical contingency plans developed for mitigating the consequences;

(iv) meetings and discussions would take place with all individuals, community representatives and representatives of specific interest groups who would be or who might be affected by the development, the purpose being in the first instance to promote understanding of the respective viewpoints and to arrive at decisions through a process of consultation which would permit the development to go forward in an orderly way. The BP personnel who took part in the many meetings of this type were the managers, supervisors and technical experts doing the work and responsible for the development. Environmental and public relations staff were used as specialist advisers, but were usually not directly involved in discussion or negotiations with outside parties.

In addition BP was very conscious of what had happened in Alaska following enactment of the 1969 National Environmental Policy Act in the USA, and decided that in future it would prepare its own environmental impact assessments for all major projects. The first of these was done for the whole Forties field development from the offshore platforms in the North Sea to the tanker terminal in the Firth of Forth.

ENVIRONMENTAL CONSIDERATIONS

The development of the Forties field production system comprised four major engineering projects:

   (i) offshore platforms;
  (ii) submarine pipelines and landfall;
 (iii) land pipeline;
 (iv) stabilisation, storage and export facilities.

Offshore Platforms

The development plan called for two platforms to be constructed simultaneously and, when completed, to be followed by a second pair. The size and weight of the platform structures dictated that the construction sites were on the seaboard, and the sites chosen by the two fabrication contractors for the first two structures presented a vivid contrast in environmental factors. At the beginning of 1972, Highland Fabricators established their base in a sparsely populated rural area at Nigg on the Cromarty Firth in Scotland, constructed a dry dock on the foreshore, and recruited and trained local labor to build the platforms. Simultaneously Laing Offshore opted for a disused existing dock on Teesside on the north-east coast of England where there was an abundance of skilled labor readily available in a densely populated area traditionally associated with heavy engineering.

Design criteria for the steel platforms required them to be anchored to the seabed by driving cylindrical steel piles to adequate resistance and grouting between each pile and its pile sleeve. Special attention was given to the potential effects of fatigue and to the dynamic response of the structures to waves. Safety factors for stress and overall stability were in accordance with regulations and included provision for a wind gust speed of 114 knots and a wave height of 94 ft, both figures based on predicted 100 year maxima. The structures were cathodically protected from time of installation with special provision for protection of the structural splash zone. The potential significance of fouling was appreciated and allowed for in the design, although there was then only limited data available for the assessment of probable rates of growth, or for the thickness and weight of the colonising marine organisms.

The design criteria to be met, the loadings to be sustained and the steel thicknesses employed necessitated the most rigorous analysis and model testing of the structures with both laboratory and field research on soil mechanics, pile driving, structure node stressing and node welding techniques.

Two platforms were installed in summer 1974, and a further two in summer 1975. Since the facilities came on stream a program of non-destructive testing and inspection by divers has been in force. During the 6 years since installation of the first structures, the program has shown that they are performing satisfactorily.

The Forties field development was supported by a series of physical and biological surveys before and during platform installation and in operation, and a biological monitoring program of the production system has been in effect since 1975. To date the biological and geochemical surveys have shown no detectable departures from the normal background hydrocarbon levels.

The gas:oil ratio of the Forties crude is low, but even so, at production rates of the order of 500 000 barrels per day the gas represents a significant quantity of energy. Gas recovery and utilisation on the Forties platforms is achieved by compressing and refrigerating the gas to produce NGL which are injected into the crude oil line for transit to shore. The residual lean gas, essentially methane, is used for driving pumps, compressors and utilities on the platforms, and a small quantity is flared.

While drilling the primary factor in maintaining control of the well at all times is to have trained and experienced personnel who conduct the operations in accordance with good drilling practice. There are, of course, back-up surface and down-hole safety devices, and once a well is in production it has automatic shutdown control for fire and over-pressure.

In the past 5 years of operation of the field, there has been no oil pollution incident arising from the drilling, production or transit facilities, reflecting the integrity and high operating standards of the system.

Submarine Pipeline and Landfall

The route survey and associated seabed investigations began in 1971, research testing and preparation continued through 1972 and pipe laying operations started in 1973. At that time the laying of a 32 inch diameter pipeline in water over 400 ft deep in North Sea conditions was outside all previous offshore pipe laying experience.

Design considerations included:

   (i) laying stresses in the pipe at the over-bend as it left the lay barge and at the sag-bend as it met the seabed;

  (ii) operating stresses and possible surge conditions that might be experienced;

 (iii) possibility of abnormal stressing of sections of the pipeline due to pipe movement or seabed change;

 (iv) possible damage due to direct impact from trawl boards or ships' anchors.

Since the pipeline was likely to operate on the seabed for a long time, the first consideration was the quality of the steel and the welding. Not only had the pipe to meet X65 quality to API standard 5XL, but in metal-lurgical terms it had to have superlative cleanliness to reduce the possibility of flaw and of welding stresses. As a further precaution a 100% weld examination was conducted.

The pipe required a high strength concrete coating that would give the best balance between the possibility of over-stressing the pipeline while laying and providing negative buoyancy and physical protection when installed on the seabed.

Most important, there had to be some basis for ensuring that it would be possible to tension a catenary of 1200 ft of large diameter pipe from the end of a barge to the seabed and successfully lay the pipe. This was achieved by exhaustive full-scale deepwater trials in the Mediterranean in the summer of 1972 before tackling the more hostile waters of the North Sea. The route for the pipeline from the field to shore, avoiding rock outcrops and severe seabed elevation changes, was chosen after extensive seabed surveys in 1971 and early 1972. In general, the route is on a seabed of predominantly soft silts and clays and this permitted about 95% of the pipeline to be buried in the seabed by high pressure jetting techniques. This degree of burial should not be regarded as normal. In the Forties case the seabed conditions were favorable for burial, but in other locations it is questionable if it is possible to successfully bury a pipeline without incurring unacceptable time and cost penalties (with no guarantee of ultimate success after numerous passes).

The landfall selected for the pipeline was at Cruden Bay and the exact point on the beach for the pipeline crossing presented an interesting environmental conflict. Landward of the beach is Cruden Bay golf course, across which a small stream flows constituting a golfing hazard. From a botanical point of view the best place for the pipeline excavation work would have been to follow the natural gap of the stream, but this would have created quite a lot of damage to the established golfing fairways. In the event, the pipeline was laid through the sand dunes to the south of the golf course. But this entailed major sand dune reinstatement (designed and supervised by Professor Ritchie of Aberdeen University). Another point of environmental interest was the way in which the engineers made use of the natural long-shore current system for disposal of the spoil from the trench where it crossed the beach and passed through the dunes. The results of the sand dune and beach work can be fairly said to have been a very significant geomorphological and botanical success and it is now virtually impossible for a person walking along the beach to spot where the line comes ashore (Fig. 1).

The pipeline is continuously monitored for possible leakage by a line balance system that compares the flow rate of the oil leaving the pipeline at the landfall and the sending flow rate from the field. In fact, there has been no leakage or disturbance to the pipeline since it was commissioned in 1975. It is inspected annually by observation from an underwater vehicle.

Although there has been no pollution of the sea from the field or the submarine pipeline there are comprehensive contingency plans for tackling any incident that might occur. The plans include immediate action by support vessels permanently stationed in the field and equipped for fire-fighting and carrying oil dispersant and equipment for contain-

Fig. 1. Landfall of British Petroleum's submarine pipeline at Cruden Bay, Aberdeenshire. Sand dune contours have been restored and vegetation has started to re-establish, some 3 months after the pipeline was winched ashore in May 1973.

ment and collection of spilled oil; if necessary additional equipment and support is available through the UK Oil Operators Association (UKOOA) emergency oil spill clean-up plan, and from the BP Group's central resources.

There was a potential conflict of interest between the offshore oil industry and traditional maritime activities. The resources most obviously at risk were the fisheries and the community that was most likely to be directly affected by our operations was the fishermen. The first meeting between representatives of the various fishing associations and the Manager, Forties Development, took place in Aberdeen on 9 February 1972, and subsequent meetings occurred at intervals throughout the development. In April 1972, the fishing representatives opined that the proposed route for the submarine line was rather close to the Buchan Deep and the Turbot Bank areas and after evaluation, BP moved the route 3 miles further north of those locations. For one of us (MML) there was nothing during the development that gave greater pleasure than establishing a rapport with eminently practical, knowledgeable and experienced Fishing Association representatives. After those Forties meetings the Fisheries and Offshore Consultative Committee was formed, bringing together representatives of government, fishing and the offshore oil industry, for consideration of matters of common interest, and in particular claims for compensation for damage to or loss of fishing gear through contact with debris on the seabed. This could have become a contentious issue because detritus on the seabed comes from many sources (two World Wars made a major contribution) with consequent problems in determining liability and effecting compensation.

The Land Pipeline

The design and construction of the 133 mile, 36 inch diameter land pipe-line from Cruden Bay to Grangemouth on the Firth of Forth did not call for new technology and research as was the case with the offshore systems, because the technical aspects of laying a large diameter land pipe-line were well established.

Environmental factors, however, were important and varied. The pipe-line was constructed along a route, predominantly agricultural, comprising 322 separately owned or leased parcels of land. The preferred route involved crossing of some of Scotland's most famous fishing rivers, the Ythan Estuary with the University of Aberdeen Field Research Station, and areas of scientific and archeological importance, most of which were already known but some (see below) were identified by company experts in the course of route selection. The route was determined in the normal way by setting out a preliminary course on maps, followed by a field survey and by walking the route and consulting with the land owners, farmers and all interested parties to determine adjustments which would satisfy all concerned without sacrificing engineering standards. Most important of all were the meetings and informal discussions with members of the public at towns and villages along the pipeline route which gave an opportunity to explain what BP was trying to do, learn of the reactions and fears of the communities near to the pipeline, and to obtain the benefit of access to the great store of relevant local knowledge. After the surveys and consultations it was concluded that construction of the landline along the proposed route would not result in any major ecological or conservation problems. This conclusion has been borne out by the operating experience of the past 5 years (Fig. 2).

It would take a book to detail all the considerations involved in route

(a)

(b)

Fig. 2. British Petroleum's North Sea oil pipeline from Cruden Bay to Grangemouth, at East Dron, Perthshire (a) after burial October 1973 (b) August 1975.

selection, construction and restoration and the following are just a few examples of the factors which were taken into account in building the landline. To avoid disturbance during construction it was arranged that the crossings of the major fishing rivers would be undertaken at times of the year which were agreed with the owners of the fishing rights at the crossing points. The route survey had revealed an area 1 mile by $\frac{1}{2}$ mile known as the Red Moss, about 8 miles south-west of Aberdeen. Examination revealed that there were a number of distinctive habitats present in this glacial feature (a kettle hole) and it was recognised by our ecologists as a site of great scientific importance. Following consultation with the Nature Conservancy it was declared a Site of Special Scientific Interest and the proposed alignment of the route was altered to move the line away from the area. Early in 1972, a survey was carried out at the Archaeological Division of the Ordnance Survey in Edinburgh, on the relationship of archeological sites and the proposed pipeline route and it was possible to avoid all the sites identified except the Roman military fortification between Bo'ness and Falkirk, where a crossing point which would cause the minimum disturbance was agreed with the archeologists.

The most interesting problem in obtaining consents to the laying of the line arose from an estate that had suffered an outbreak of brucellosis. The land was being sold and when the purchaser learned of the brucellosis he withdrew from the proposed contract. The problem was to find someone who acknowledged ownership, since the vendor reckoned he had sold, while the erstwhile purchaser reckoned he had not purchased. In the event we avoided delay by buying the land, and both there and elsewhere special care was taken to ensure that there were no livestock movements between holdings along the pipeline right of way.

Work on the landline brought a large part of the population of the east side of Scotland into contact with North Sea oil development for the first time. Most people showed great interest in what was happening and it was only reasonable that they be kept up-to-date on the progress of the Forties development. Hence the *Forties News* came into being: it was an information newspaper first published in 1972 and produced throughout the life of the project. It was distributed to the public through a variety of agencies, including local societies and shops in the different towns, and was supported by a mobile trailer-borne exhibition, showing the planned Forties development and the techniques of offshore oil exploration and production, which toured all of Scotland.

The landline is completely buried throughout its length except at valve stations, and has operated satisfactorily and without disturbance to the environment. As with the offshore systems there are contingency plans for dealing with unforeseen incidents.

Stabilisation, Storage and Export Facilities

The landline ends at Grangemouth refinery where there is provision for further oil and gas separation to stabilise the crude oil for storage and export. Separation and gas processing facilities are standard.

The decision to establish a loading terminal in the Firth of Forth gave rise to two important decisions on the siting of facilities. First, a location for a tanker loading terminal had to be found that would meet the requirements of adequate water depth in a protected anchorage, access to a navigable channel, suitable seabed conditions for piling and which would not obstruct other shipping in the estuary. The place chosen was $1\frac{1}{4}$ miles east of the Forth Railway Bridge and 2500 ft off the south bank of the Forth. Secondly, oil storage and de-ballasting tanks had to be constructed on a site close to the south bank of the Forth to take the Forties stabilised crude from Grangemouth, to deliver crude oil to the tanker berth, and to process ballast water from tankers. The location selected was at an oil shale mining site near Dalmeny, immediately south of the A90 road and about $1\frac{1}{2}$ miles from the south shore of the Forth.

The most important environmental consideration relating to the tanker terminal is the risk of oil spillage during loading operations or from an accident to a tanker in the estuary. In the event of a pollution incident occurring at certain times of the year important seabird populations could be at risk. Since the commissioning of the terminal in January 1976, prevention of pollution during berthing, unberthing, or loading, has been a prime objective of all concerned. The safety codes and good operating practices derived from BP's long experience of tanker operations worldwide are strictly applied, and while it is no cause for complacency, it is worth mentioning that since the first tanker, *British Commerce*, loaded in January 1976, over 800 tankers have been loaded without a significant pollution incident.

BP's oil spill contingency plan for the estuary is integrated with the Clearwater Forth plan which coordinates the oil pollution control activities of industry, local authorities bordering the Forth and the concerned agencies of central government.

Although the site proposed for the tank farm was at an old shale mining site, it was within the Green Belt of Edinburgh and BP's primary environmental concern was the visual impact of large tanks in the Green Belt and close to a main approach road to Edinburgh. In consultation with local landscape architects BP proposed demolition of the shale bing (tip) to provide material for a naturally contoured protective bund for the tank farm. This was acceptable to the authorities and the tank farm development at Dalmeny is so well camouflaged that it can only be seen from the air (Fig. 3). The quality of the landscaping is evidenced by the fact that the tank farm received a European Architectural Heritage Year award. However, attitudes change quickly and there is now a movement to preserve old shale bings in that part of Scotland as important industrial archeological sites.

Fig. 3. British Petroleum's crude oil storage installation, Dalmeny, Scotland.

CONCLUSION

This rapid and highly selective summary can only give a flavor of the environmental factors which had to be taken into account in the development of the Forties field. There are two other important matters which warrant some mention here.

First, it is frequently asked how much the environmental protection measures cost. It is doubtful if a sum of money can be ascribed to this aspect of the project and the question misses the point: good environmental practices are now an intrinsic part of the design and implementation of major projects; they are not cosmetic 'add-ons' which can be readily costed for public relations purposes.

Second is the impact of the influx of oil construction and operational workers on established communities. In our opinion it is too soon to satisfactorily and objectively assess the social consequences of oil development in a local or regional context.

Sir Nevil MacReady in his presidential address to the Institute of Petroleum in July 1980 stated that surveys had shown that the public in general did not think the oil companies keep them informed, that they do not believe most of what oil companies say and that they consider the companies secretive and selective in the facts that they make available. *Plus ca change* . . . and it is against this background that the environmental conservation factors relevant to oil-related developments have to be evaluated. Reasonable decisions made or compromises effected on the basis of good data can and should be made widely known and it would seem worth while for the oil industry to make greater efforts to inform the public about the way in which it sets about solving environmental problems. Both sides can benefit from an active process of peer review which eschews the emotional in favor of honest and forthright debate.

9

The Problem of Setting Standards for Environmental Protection: The United Kingdom and Norwegian Approaches

D. W. FISCHER

Institute of Industrial Economics, Bergen, Norway

ABSTRACT

UK and Norwegian experiences in setting standards for oil pollution in the North Sea are compared. Issues include the relevant bodies setting such standards, the bodies who would be expected to be involved but are not, the information base and links, the actual standards set, their effectiveness and the analyses used. The paper notes the conceptual framework used in research on the above issues. Areas of convergence and divergence between the patterns of UK and Norwegian decision-making are noted where the standards themselves converge while the process for setting them has diverged.

INTRODUCTION

P. T. Barnum, the American circus entrepreneur, once noted that a lamb can co-exist with a lion if one has a large reserve of lambs. This statement aptly points out a dilemma in the setting of constraints on energy developments for protecting the environment. Is the environment equivalent to the lamb and energy to the lion or vice versa? Does environmental protection via high standards threaten to hamper energy development in an energy-short world, or does energy development via low standards threaten the environment with increasing pollution and degradation? Striving to keep a higher supply of environmental quality in the same arena as high energy demands and striving for high levels of energy development along with high environmental quality demands require difficult sociàl trade-offs between them. This is a value-laden problem; this paper starts from the actual position of the UK and Norway where they attempt to balance energy development, or more specifically North Sea oil, with a regard for a quality environment.

While oil kills some forms of life upon exposure, its continued presence in the environment may kill only as a result of a complex and insidious causal process. The first link in the chain is clear: some natural and refined oil fractions are highly toxic to some forms of life. Thereafter the links are less clear and it cannot be easily proven that the environment is significantly harmed through oil spills. Continuing oil pollution from all sources is in the vicinity of 5×10^6 tons per year worldwide whereas continuing pollution from oil drilling and production is around 50 000 tons. Accidents, however, can cause oil pollution to occur in dramatically large amounts such as at Bravo where 22 000 tons were lost. Knowledge of actual risk levels for various populations and their habitats is fragmentary and sometimes circumstantial. The environmental significance of the magnitudes and rates of oil spills in the North Sea seems elusive at present. With such a lack of known risk levels for life expectancies in the environment, empiricism reigns. It is the empirical approach of protecting the environment from oil discharges that is briefly examined in this paper.

It is of interest to note that North Sea oil was discovered about the same time that the UK began to reorganise its governmental machinery to control environmental pollution, followed by Norway in its creation of similar machinery. Thereafter oil and environmental policies evolved in both countries in a co-existence that recognised certain rights of each. This paper is focused on the machinery that UK and Norwegian bureaucracies employ in oil–environment questions, in particular the setting of standards for environmental protection in relation to discharges of North Sea oil on land and sea. While standards are only one of several environmental protection tools, they have emerged as a central device used in the regulatory processes at the interface between oil discharges and the environment.

GROUPS INVOLVED IN STANDARD-SETTING

Any standard is a specific statement that has been evolved from the interplay of a standard-setting process. Environmentalists and oil interests equally find fault with standards. Researchers used by both also criticise standards. Wherein is 'truth'? The truth is that a standard is not based on a precise scientific statement of cause and effect but rather on the outcome of the interplay among many groups. The regulatory procedures for establishing standards fulfil a mediating role between environment and oil interests so that the resulting standards rarely meet the objectives of single interests. Because of the non-scientific character of a standard the process for setting it becomes just as important as the standard itself. In particular, the underlying reasons for the difficulties in setting standards can be attributed to:

(i) uncertainty about actual effects of oil spills in the environment as well as on human well-being; experts differ considerably in their estimates and often are unable to make precise judgements;

(ii) the regulator being more than a single decision-making unit with various administrative and expert groups involved to different degrees both formally and informally; in addition there are certain affected groups who perhaps ought to be included but are not;

(iii) conflicting objectives such as environmental, economic, energy, engineering and political as well as varying legislative and administrative bases; these objectives are difficult to specify and few are commensurable;

(iv) the effects of the pollution and the standards set being distributed unevenly and often imperceptively among groups and over time; it is difficult to make trade-offs explicit for understanding the gains and losses.

We can suggest a reasonable approach to setting oil discharge standards that can account for most of these difficulties. We focus on the various groups expected to be interested in the process of standard-setting and determine the groups that should interact in setting either discharge and/or environmental standards. The premise of this approach is that the more uncertainty that exists the greater the necessity for a comprehensive and integrated system linking all groups likely to be affected by whatever the outcome. Such an approach would seem to account more effectively for balancing productivity, efficiency, need, equity, risk and quality in the setting of standards than an approach based on single separate interests. Ideas of fairness in relative terms tend to dominate more than in real terms (as with wages). Certain questions are posed:

(i) which groups are included, excluded and why?

(ii) which objectives are considered?
(iii) which alternatives are considered?
(iv) what information is gathered and evaluated?
(v)   how are the above meshed to set standards?

Based on the answers to these questions standard-setting practises can be evaluated and eventually compared between countries. This approach was applied to both the UK and Norway for setting oil discharge standards, especially in the offshore. There are five basic groups: regulators, developers, experts, impactees and exogenous or external groups. Also, the machinery or organisational linkages among them are important for their interactions (or lack of) in the process of structuring objectives and alternatives and generating information to set standards.

OFFSHORE STANDARD-SETTING

What emerged from some of our work can be shown in highly reduced form where the process of setting offshore oil discharge standards for platforms can be seen. The major differences in the process between the UK and Norway can be seen in the placing of the regulator and the degree of openness with giving and receiving information. The UK regulator is the Petroleum Engineering Division (formerly called Production) within the Department of Energy while the Norwegian regulator is the independent State Pollution Control Authority separate from the Department of Environment. While it would seem that the UK would opt for greater openness with information, given its regulatory machinery within the very unit most interested in oil production, the UK relies on a doctrine of collective responsibility whereby one official speaks for all interests within the public sector and therefore does not openly share and balance information with those outside. In Norway openness is professed but is not pursued, and indeed little new information has been gained from impactees by practising it. In addition, Norway views oil discharges pessimistically in both amounts and impacts, partly from its interests in maintaining fisheries exports and their traditional way of life. The UK views oil discharges far more optimistically, partly from its interest in more rapid oil production and in having less of a fishery constituency. Finally, UK membership in the EEC was a basic motive for generating the UK standards in compliance with EEC directives, while Norway's basic motive sprang from its concern for its traditional patterns.

Despite these differences source of regulation, source of expert information and attitude have not affected greatly the actual standards set as both countries have created similar offshore oil discharge standards (30–55 ppm) based on existing availability of pollution control technology. Is our suggested approach then of little value?

SHORTCOMINGS OF THE STANDARD-SETTING PROCESS

While not shedding light on the offshore standards generated by the process as described, our approach does highlight where we could expect conflicts should damages from oil pollution be perceived and where we could improve the effectiveness of the process.

Developers express concern over the apparent lack of interest in their sponsored research and views on the degree of severity of the standards. Such concerns can lead to conflict over questions of liability, relationships with the regulator, and future research and linkages for pollution control technology alternatives and their integration into platform design.

Regulators lack information with which to judge proposals from developers, especially on alternative environmental technologies, impacts and economic costs. This lack can lead to higher costs for developers, consumers and society at large through higher prices, more expensive methods, displaced activities and delays from repeating research already done elsewhere. It also leads to a lack of environmental information being available as a basis for setting the standard.

Independent experts lack regular access to regulators and developers so that they are excluded and may use other ways to bring forward their views. This process heightens professional and public criticism and leads to more resistance to development proposals. It also generates confusion and reduces the experiences of experts in making probability judgements.

Potential sufferers from offshore oil pollution include fishermen, consumers of fish, coastal residents, tourists, tourism industry, local governments, conservationists and environmentalists. Excluding such impactees from standard-setting abets feelings of non-involvement, opposition and misinformation.

Exogenous groups can act to constrain or further oil development via standard-setting procedures. The roles of international bodies as well as other parts of the central government and industry are not well integrated into generating alternatives for standard-setting.

Thus the offshore oil pollution standard-setting case in the UK and Norway suggests that conflicts can be expected in information, leading to higher costs for all groups concerned from the lack of participation, lack of other objectives and alternatives being considered, and lack of information priorities being established.

ONSHORE STANDARD-SETTING

A key feature of the offshore is a lack of experience since few groups can identify themselves with it. No one lives there permanently and direct links between oil pollution and an offending platform cannot be easily traced. Onshore neither of these aspects are true: permanent residents and others readily identify themselves with onshore areas and sources of oil spills and resulting damage is often easy to detect on and close to the shore. Thus on land we can expect to find greater access by experts and impactees, greater awareness of other objectives, wider interest in alternatives and more environmental information for assessing impacts.

In comparing offshore and onshore standard-setting processes, the major differences are the addition of Scottish and local government machinery plus the presence of other advisory experts in the UK. Regulatory authority in the UK onshore is now delegated to local authorities. Central government directives can constrain or further local standards for oil development proposals. No formal means exist for eliciting environmental information unless the proposal involves a fundamental change in land-use possibly necessitating a local public inquiry; however, non-residents, environmentalists and other experts are included only by invitation. The Scottish Development Department (SDD) advises local governments through its various units, including its pollution inspectorate. However, environmental bodies must rely on an advisory role, and none has direct executive authority over oil projects.

Lack of formal executive authority for environmental bodies in the UK has led to creation of new organisational form: the reticulist or network role. The Central Unit for Environmental Pollution, the Nature Conservancy Council in Scotland, the SDD and the Commission on Energy and the Environment all act in a network role to provide advice and act as a broker between bodies with executive authority. Their roles in standard-setting are informal, closed to outsiders, and are in the background.

In Norway the SPCA maintains a strong presence in the onshore since its powers include all of Norway. Any oil project is subjected to the

SPCA's pollution standards which are also accepted by local governments. Formal links are used in tying the environmental system to oil projects and the SPCA can exercise veto authority (which it has yet to do). Both executive authority and expertise are within a single organisation and ties are maintained with other experts and selected impactees. However, a myopic view of research results from elsewhere is maintained and these standards were belatedly applied at the first large onshore project at Rafnes.

How does the UK decentralised development authority with informal environmental roles compare with the Norwegian centralised and formal environmental authority? There is apparently no significant difference in standards; however, in the process of standard-setting and critiques of it differences are significant. The UK approach has far more national critics, particularly from independent experts and 'professional citizen' environmentalist groups. Specific standards in both countries depend on available technologies and yet neither country supports extensive pollution control research into alternatives, instead relying on entrepreneurial endeavor and oil company integration of it. Both countries rely on discharge standards tied to specific technologies as opposed to ambient standards tied to environmental capacities; neither applies trade-off analyses among polluters whereby costs of compliance can be reduced among polluters and regulators. Finally, neither country looks extensively at environmental criteria when compared to technological criteria but nor do other countries.

CONCLUSIONS

There is convergence in the kinds of standards set for oil-related discharges but a divergence in the process of setting such standards. This paper has stressed institutional arrangements among groups rather than actual standards because organised relationships among groups determine the kind, quality and timeliness of information and decisions as well as an ability to adapt to changing conditions and hence reconcile conflicting interests among environmental and energy programs, among others.

The following lines of endeavor may be profitable in future:

(i) increased attention to the merits of allowing participation by a wider array of publics: independent experts and impactees, local, national and international;

(ii) increased information on environmental impacts and capacities from increased national research, acceptance of research findings from other sources, and dissemination of such results to 'outsiders';
(iii) increased focus on assessing alternatives both in control technologies and in other means of control such as emission fees;
(iv) increased concern for reducing net social costs for pollution control to developers, regulators and consumers through assessment of alternative control strategies or processes.

In pollution control and environmental protection generally the best is often the enemy of the good. Given limited regulatory, fiscal, environmental and energy resources control strategies must be selective. It appears that the lamb–lion roles are selective. The environmental lamb can lie down beside the energy lion without being defiled, and vice versa, depending on available resources and needs. The process for protecting the lamb leads to its standard of protection, and the amout of lamb reserve necessary regardless of whether the environment is the lamb or the lion.

BIBLIOGRAPHY

Berrefjord, O. (1979. *The Rafnes case: a study on decision-making.* University of Bergen, Institute for Sociology, Norway. (Mimeo.)

Central Unit for Environmental Pollution (1976). *The separation of oil from water for North Sea oil operations.* Pollution Paper No. 6. HMSO, London, UK.

Department of Environment (1976). *Environment impact analysis.* Research Report II (Catlow and Thirlwall Report). DOE, London, UK.

Fairclough, A. J. (1976). *The UK approach to environmental matters.* Central Unit for Environmental Pollution, London, UK. (Mimeo.)

Fairclough, A. J. (1978). North Sea oil hazards. *The environment interface and implications for management.* Central Unit for Environmental Pollution, London, UK. (Mimeo.)

Fischer, D. W. (1979). Policy issues in standard-setting: a case study of North Sea offshore oil pollution. In *Energy risk management*, ed. G. Goodman & W. Rowe. Academic Press, London, UK, pp. 61–85.

Fischer, D. W. (in press). *The North Sea oil–environment interface.* Bergen Universitetsforlaget, Norway.

Fischer, D. W. & Winterfeldt, D. v. (1978). Setting standards for chronic oil discharges in the North Sea. *Journal of Environmental Management,* 7, 177–99.

Fitzmaurice, V. E. (1978). Liability for North Sea oil pollution. *Marine Policy,* April 1978, 105–111.

Ministry of Environment (1975–76). *On pollution control measures.* Parliamentary Report No. 44. Oslo, Norway.

National Academy of Sciences (1977). *Implications of environmental regulations for energy production and consumption.* NAS, Washington, D.C., USA.
Read, A. D. (1978). *A regulator's view of the standard-setting problems for oil water discharges from North Sea offshore oil installations.* Department of Energy, London, UK. (Mimeo.)
Scottish Development Department (1977). *Oil, gas and petrochemicals.* Planning Information Notes. SDD, Edinburgh, UK.
Winterfeldt, D. v. (1978). A decision aiding system to improve the environmental standard-setting process. *Systems analysis applications to complex programs,* ed. K. Chihochi & A. Straszak. Pergamon Press, Oxford, UK.

10

The UK Sector: Forecasting the Effects of Production and Consumption

F. H. MANN
Bank of Scotland, Edinburgh, UK

ABSTRACT

A personal, subjective assessment of the desirable position of oil in the UK's energy plan indicates that the impact of oil on the environment cannot be considered separately from that of all energy alternatives, which in the UK are principally reduction of industrial effort, conservation, nuclear power, coal, gas and oil. Each has a contribution which it could make towards fuelling Britain in the short- and medium-term, and each would incur its own environmental costs, which in the case of oil are relatively acceptable. Given that oil will be used extensively in the UK, we have the choice, to some extent, of importing or producing it. Each of these choices also has its environmental costs. It is argued that indigenous production is normally preferable to imports on environmental grounds. Moreover, energy use and production in the UK is a small part of energy use and production worldwide. International dependence on oil and international effort to supply it impinges on the UK and the UK must accept its international responsibilities if it is to make the most of oil.

INTRODUCTION

This paper is written not by an expert on environmental matters but by an ordinary member of society who sees the experts concentrating on details of environmental problems while apparently neglecting more important aspects of life, i.e. not seeing the wood for the trees. Considering the alternatives, I believe that oil is environmentally benign.

The main problems of pollution are due to the consumer, not the producer. In fact, the degree of damage to the environment steadily increases down the chain of oil use from production to consumption. This concept is not supported by the emphasis of this conference, in which titles of papers strongly suggest that the principal environmental problems are upstream in the production and bulk transport sectors. Few papers are concerned with the refining and processing sectors, and none with product distribution and the burning of fuel.

Certainly headlines are made by big blow-outs, such as those at Ixtoc and Ekofisk. Yet I suspect that very few of us have even spoken to people who have seen blow-out damage. In the bulk transport sector, tanker disasters, like blow-outs, are also good for headlines. However, such events are relatively rare and tankers account for relatively little pollution. Much of our beach pollution is due to oil users, for example, cleaning bunkers.

Further downstream, refineries, processing plants and their associated storage tanks are relatively prominent visually. They use quite a lot of countryside, coastline and water, and affect air quality, if only slightly. The use of oil, however, really becomes noticeable as it is distributed. Every day each of us sees, hears and smells road tankers, service stations and other distribution facilities.

Finally, ignoring the relatively small non-fuel use of oil, the fuel is burnt. All the other upstream environmental aspects of oil are relatively unnoticeable compared with petrol fumes, exhaust emissions and chimney products which impinge on our senses every day and often blanket millions in the acute misery of smog. Our streets are stained with oil spills and our rivers contaminated with oil swept down from sewers.

As for oil so, to a large extent, for other fuels. Hence in rating fuels environmentally I concentrate on the downstream end.

TRENDS IN OIL USE

Variations in the UK's total primary fuel input over the decade 1969–1979 have been remarkably small. Effectively, our fuel input has been constant at $84 \pm 4 \times 10^9$ therms each year, although the immediate trend is sharply downwards, presumably because of the recession. However, in spite of this constancy of fuel input, the pattern of fuel use has changed considerably. In particular, oil's share of the primary fuel input has declined from 43% to 40%, although transport's share of the end use of fuel has risen from 19% to 23%. Since virtually all transport fuel comes from oil, for which there is no serious competitor in sight, transport fuel use sets a lower limit to oil consumption in the foreseeable future: the sharp fall in oil consumption experienced over the last year cannot continue indefinitely (if oil use were to continue to decline at the same fractional rate as recently, the oil reserves already proved in the UK North Sea would exceed the UK's total remaining requirements for oil).

Thus the general impression is that oil use in the UK is declining despite our North Sea finds and I see no reason why this decline should naturally reverse. Hence I foresee a decreasing impact environmentally in that part of the oil use chain from refining/processing through distribution and consumption. Of course, more crackers will be built as transport fuels become more and more expensive relative to residual fuels, but only the refiners are likely to notice it. Upstream, changes will take place offshore as new fields are brought into production, probably relatively many small new fields replacing the big ones already well into their total life. However, the total environmental effect will be small mostly at onshore construction and supply sites, and not very noticeable even there.

It is worth noting that a recent world coal study[1] foresees similar oil use effects throughout the Organisation for Economic Cooperation and Development (OECD) countries, the world's main consumers of oil. Outside OECD, however, catching up in consumption by the rest of the world is expected to result in a slowly rising total consumption through the 1990s at least.

POSSIBLE REDUCTION IN ENERGY USE

The environmental impact of oil could be reduced by reducing oil consumption and an obvious way of doing this is to reduce total energy consumption. This might be achieved either by conservation, i.e. more efficient use of energy, or by reduced application of energy.

Of course energy conservation must be effected. It has already caused some reduction in UK energy use but much more can be done. While conservation is worth encouraging in all ways, energy price rises will probably bring it about nearly as quickly as we could wish.

Reduction in the application of energy, essentially reduction in industrial effort, is a policy for which I have no sympathy and do not consider as a serious possibility.

USE OF OIL SUBSTITUTION

A useful reduction in oil's environmental impact could be forced by substituting for oil some fuel with a lesser impact.

Oil has several uses, principally for electricity generation, industrial heat, residential heating, transport fuel and petrochemical feedstock. Substitution for transport and petrochemical purposes essentially requires the preparation of synthetic oil. Use of synthetics would not eliminate refineries, processing plants, the product distribution chain nor the consumer, so the worst part of oil's environmental impact would remain. The proposed base for synthetic oil is coal. Hence oil platforms, pipelines and tankers would be replaced by coal fields and synthetic oil plants. Environmentally this would scarcely be an advantage.

For industrial and residential heating, oil conceivably might be replaced by electricity, natural gas or coal. Conversion of primary energy to usable electrical heat is inefficient and militates against a major shift from other fuels to electricity for most industrial and residential heating. As a primary energy source for electricity generation oil competes with natural gas, nuclear energy, coal and the so-called exotics (wind, waves, tides, geothermal, solar, biomass). The exotics are unlikely to compete with oil in the UK in this century. Technical reasons alone relegate these options to less than 5% of the primary fuel supply and other considerations make even this level unlikely. So from the environmental point of view oil has only three significant competitors in the medium-term—natural gas, nuclear energy and coal.

OIL'S COMPETITORS FOR FUELLING THE UK

Natural Gas

Natural gas is less damaging to the environment than oil, coal or nuclear energy. It is clean burning, requires a minimum of equipment and produces no solid or liquid waste. It does not need huge storage facilities. It is silently and almost invisibly transported and distributed in bulk, does not contaminate if it leaks, makes no demand on water supplies and needs very little processing. It requires small production facilities relative to the energy produced. Gas can and does cause fires and explosions, but any fuel mishandled can do this.

Gas really has only two drawbacks as a fuel. First, the preferred distribution system makes it available only to concentrations of fixed installations, e.g. in towns. Secondly, supplies are inadequate for all possible applications. In view of its advantages and its effective shortage it is under-priced. To pay for it at its real value would give us a shock comparable to the one given by OPEC when we were asked to pay for oil at a price reflecting its value.

Nuclear

Nuclear energy is in practice limited to electricity generation. Theoretically it could supply industrial heat directly but a sufficiently big customer has yet to appear. Eventually it may be used for synthetic gas generation, but that day is still far distant.

The environmental aspects of using nuclear energy in the UK are as follows: bulk transport of raw material requires some use of ships and thus of coastline, but does not cause significant coastline pollution; refining and processing of raw material requires plant at least comparable to that for oil. Refined fuel storage, however, is negligible compared to that for oil. Nuclear power stations are relatively bulky and use appreciable amounts of water. Waste products have to be disposed of, but that is really only a minor problem. Atmospherically it is non-polluting. It does provide a risk of air, water and land contamination although that is a risk rather than a reality. Paradoxically, although the probability of contamination is extremely small the fear of it is very real and perhaps constitutes the biggest problem. Clearly nuclear power is not as attractive as natural gas, but it is not bad. Considering only the electricity generation to which it is limited, environmentally it seems to be on a par with oil.

Coal

Transport could return to the steam age and coal, but the thought of a land full of coal-fired trains, trucks and buses is too horrifying to consider further. Of the four options, coal is the least attractive fuel for power generation, industrial and residential heat.

Its drawbacks include massive bulk storage requirements. Its distribution demands more storage and more transport than oil, and both are pretty filthy. Burning it either produces atmospheric pollution or calls for major supplies of absorbents to scrub exhaust gases. There is a major by-product disposal problem (pit spoil and ash) and water is used in processing. In fact it is hard to find anything good environmentally to say about coal production.

However, coal itself poses little long-term contamination risk, and the UK has relatively secure supplies. Security is a significant factor in considering preservation of our environment.

PREFERRED FUEL USAGE

Based on these arguments, gas must be the fuel preferred for fixed plants and for town distribution. Oil is the preferred fuel for transport and for the provision of heat and power at relatively small installations in remote locations. However, gas supply is limited and gas should be treated as a premium fuel, whether it is priced as one or not. The obvious partial substitute for oil is nuclear heat for electric power generation, and in order to maintain security of power supplies, coal use must continue. Since oil lost its price advantage over other fuels there has been a strong tendency to replace it, except for petrochemical feedstock and transport for which it really has no competitors. This tendency, which is environmentally attractive, is likely to continue with the installation of more and more cracking, leaving less and less heavy fuel for power generation and industrial heat. However, fuel oil cannot be backed out of use overnight and the environmental advantages of its elimination do not make desirable a major effort to speed up the process. Hence, we shall continue to use large amounts of oil for some considerable time.

OIL—PRODUCE OR IMPORT?

The environmental arguments for importation or home production tend to hinge on the fact that home production emphasises the use of pipelines whereas imports emphasise the use of tankers. Pipes have major advantages in minimising the use of coastline, minimising coastline pollution, lessening oil spills generally, and lessening land-use for storage (because supply is more continuous). Pipes also improve security of supply, something which tends to be forgotten between Middle East wars and OPEC cut-backs. But both actual and potential broken supply do lead to greater use of land and facilities.

Against home production is the minor disadvantage of oil production installations, which even on land are hardly noticeable, and onshore support activity. Risk of pollution from spills is more a risk than reality, as Ixtoc and Ekofisk showed.

To me the supporting industrial activity is a positive advantage (environmentally as well as economically), oil production plants are attractive, and everything is in favor of our own production. Eventually, of course, it may be necessary to import, but this is probably best put off for as long as possible.

Those who find industrial activity unattractive and oil wells hideous should note that so long as we are prepared to use oil and oil products we should help the world to produce them. Every little helps, as is shown by the volatility of oil prices with small fluctuations in supply. Moreover the gain to the world as a whole is not just immediate access to oil but also development of ideas and techniques which will improve oil production in the future.

In addition, if home oil production is not pursued as vigorously as possible, rising oil prices will bring action elsewhere in the world, and the national vigor and expansion which seems to accompany oil production will go to another country. Such national vigor and expansion will probably do more to make our environment attractive than many of the physical conditions discussed above.

CONCLUSIONS

The environmental impact of oil will decline because declining world oil supply and the resulting increasing cost of oil will ensure that we use less of it whether we want to or not. However, the adverse impact of oil on our environment can and should be reduced by substituting environmentally more attractive natural gas for oil to the limited extent possible.

At present, complete replacement of oil by environmentally more attractive alternatives is not possible. This is not just because of costs: the possible total oil replacements are less environmentally attractive than oil. Home production of oil is desirable, partly because piped oil supplies are environmentally attractive compared with tanker-borne imports but also for reasons of security of supply, industrial vigor, and to play our part in supplying the world with energy.

REFERENCE

1. Wilson, C. L. (Project Director) (1980). *The world coal study*. Ballinger Publishing Co., Cambridge, Mass., USA.

Discussion

F. P. Tindall (Lothian Regional Council, UK). There is great value in having sites designated in advance. Mr Carder admitted this was the crucial factor in his Company selecting Mossmorran. The national planning guidelines have been developed in the Forth Estuary, which includes the Mossmorran site, by the three regional councils working together with central government to identify the possible sites and service requirements for oil-related development and to understand industrial hazards arising from them. The Forth Estuary study's[1] objective was to identify the likely sites where the infrastructure services could be provided and that necessary pipeline routes could be safeguarded. The study involved not only environmental impact analysis but gathering all the necessary land-use information and service provisions. The sites identified as a result of the study will be incorporated in structure plans and local plans. Thus the local authorities will have taken a positive view of where oil-related developments should be and where they should not be. Such a study is a step in overcoming the difficulties of environmental impact analysis which relate to individual sites rather than a whole range of possible sites.

In my view it is the local authorities who should commission and steer environmental impact analysis rather than the industrial developers, although they should meet the costs. This is particularly important in view of the great scepticism which greeted some of the environmental impact statements concerning the Mossmorran development.

[1] *Forth Estuary study* (1979). Prepared by Central Region, Fife Region and Lothian Region in association with the Scottish Development Department, Edinburgh.

J. Carder. Shell feels that impact studies made for Mossmorran have been adequate and envisages none further there. However, oil companies can only welcome independent study and indeed anything which contributes to solution of environmental problems.

P. Daniel (Department of Architecture, Edinburgh University, UK). Is there some means for the records now being made on the impact of existing oil wells on the ecology of sea life to be made available internationally? Most countries involved in oil development must have similar problems which could only benefit from an exchange of information.

A. B. Viig (State Pollution Control Authority, Oslo, Norway). The Paris Convention (see Marstrander, these proceedings) has a working group on oil pollution, through which results of the studies are made internationally available.

Anon. Acquisition of land in Scotland for oil development is levelling off and we are now more concerned with processes. What sort of monitoring, particularly atmospheric, does Shell plan, particularly for the Mossmorran site?

J. Carder. Various safeguards have been imposed by the Secretary of State in the outline planning consent for Mossmorran. These conditions are very strict. During construction risk of atmospheric pollution is very slight, being confined to a small amount of blasting. During operation site management will, of course, adhere strictly to the limits laid down for noise, atmospheric pollution and so on.

J. Milligan (Strathclyde University, UK). Shell's estimate of 60 000–80 000 additional jobs from oil-related development in the 1980s may be wrong, and there is little basis for such a forecast. Simulation modelling is one way to combat such uncertainty. Is information from companies freely available to government and is the response of government to bring experts and local authorities together? The possibilities can then be covered in structure plans. The Wheatley Commission[2] has stated that Scottish regional governments should be informed of company plans likely to have significant impact. How does Shell view this problem, which concerns the confidentiality of business?

[2]Royal Commission on Local Government in Scotland (1969). Command Paper 4150. HMSO, London, UK.

J. Carder. In development of any project like Mossmorran, discussion and cooperation with local government is absolutely essential, and no large-scale development can possibly proceed without it. Regional government would be consulted on any future plans before those plans advanced beyond a critical point.

K. Alexander. There are three points worth emphasising.

First, human error is a risk one can never avoid, and the way to minimise it lies in good management and training. Those are very important parts of any company's environmental protection activity.

Secondly, the company must recognise that communications are about feedback, not simply transmitting information to people. Companies must be flexible enough to adjust to feedback, whatever form it may take. This point was made in regard to complaints procedure, and it is obviously more widely applicable in the early stages of any development.

Thirdly, in north Norway, the opportunity for better environmental planning would alone have justified the long lead time from which economic planning also gained advantage. There were environmental problems which might not have been overcome in a shorter time. This may be even more important if local people in sparsely populated areas distant from the main centers of industry and commerce are to have an opportunity to be employed in such development projects.

A. L. Walker (Department of Management Studies, Glasgow University, UK). Some members of the fishing industry are concerned about the amount of industrial garbage left on the seabed by the oil industry in general. What steps would BP take to alleviate this problem?

M. M. Linning. In general fishermen have a reasonable case about the amount of debris, particularly from construction barges, being dropped on the seabed near platforms, pipelines, etc., which gets snagged in nets and so on. As a result, the Fisheries Consultative Committee was formed under the aegis of the Ministry of Agriculture, Fisheries and Food, with the oil industry association UKOOA and fishing representatives. They set up a scheme of examination of incidents where debris had interfered with trawling. The exact conditions are complex but this system is working reasonably well, albeit against a background of continuing complaints that there is still too much debris.

There was a lot of debris in the North Sea, e.g. old wrecks, which attracted no attention until oil operations started in deeper water.

H. Moggridge (President, The Landscape Institute, UK). Although it is Dr Linning's view that consultation with the public should be done by managers only, it may be helpful for others, e.g. environmental consultants, to be involved in consultation. Restricting consultation to managers may contribute to industry's reputation for secrecy.

M. M. Linning. Environmentalists are one of a vast spectrum of specialists involved in an industrial development who must be carefully coordinated. Environmental matters are not peripheral but should be an integral part of development. The only way to integrate them is to make managers answerable for them, and not allow managers to shrug them off to another individual.

F. G. Larminie. BP's consultations were conducted by the engineers and managers on the project, but they did have technical experts available when required. It is not a good idea to put in public relations men and environmental experts just because that is what the public want to hear, because their limited knowledge of engineering is quickly exposed by detailed and awkward questions. When considering crossing points of the main salmon rivers, for example, it was essential that experts were available to back up the management's engineering comments. The importance of the whole exercise was not just the telling but that it quickly became a dialogue between concerned parties, and we learned much from it.

H. Moggridge. In conditions where my brief has been liberal and allowed me to communicate with the public freely, I have been able to give a better service through more intimate knowledge of people's problems of land-use and lanscape.

F. G. Larminie. In addition to general meetings about the whole project wich generated an enormous variety of questions and issues, BP had numerous independent and detailed environmental discussions where this was identified as the subject of the meeting, and in this way people were given the opportunity to debate the specifically environmental issues.

Anon. Better understanding of the marine environment requires more research on food chains, reactions to physical conditions, etc. In order to predict the effects of perturbations it is vital, in the absence of those perturbations, to increase knowledge of ecosystem function.

In terms of Mr Mann's concern with downstream impacts we have, in dealing with atmospheric pollution, been obsessed with dealing with sulfur dioxide. Pollution from oil, however, is much concerned with the oxides of nitrogen, which largely depend on the temperature of combustion.

There is at present much interest in trying to set environmental standards, but in reality the knowledge we have to do it is at an abysmal level. We really are at square one.

J. McCarthy (author). Professor Fischer indicated that public participation in the oil scene in the UK was rather less than he expected. The UK's system of public inquiries, however, is at least one public contact point, albeit a limited one. Is there anything comparable in Norway?

D. W. Fischer. The public inquiry system in the UK concerns changes in land-use, not in setting the standards, but I agree that it is too limited. Changes in land-use often affect parts of the country much beyond the designated region over which the local authority exercises control, even if only perceptually, so that there ought to be some broader avenue of approach. In Norway there is no formal public inquiry system. There is a building act which allows neighbors to have some say on impacts potentially affecting them. There is also a new comprehensive regional planning act that will allow public hearings but no regulations are yet written. So there is a degree of involvement, the effectiveness of which will be tested if gas is landed on the west coast of Norway.

In the standard-setting process, bodies that tend to be excluded are experts in universities or independent institutes, certain bodies of international expertise and certain international environmentalist bodies that in some but not all cases can give better information. Including them would mean fewer problems in the future.

W. J. Cairns (W. J. Cairns and Partners, Edinburgh, UK). Environmental impact assessment (EIA) is a process in which various environmental parameters are measured and compared with standards. We are, of course, able to make increasingly sensitive or fine grained measurements of environmental factors. Medical science provides good examples.

Perhaps EIA should be used in a more imaginative way. At present it is too often regarded as an activity which dictates and is the enemy of economic development. While EIA may sometimes be abused in the UK and abroad both the environmentalist and the developer should be concerned in making the process more sophisticated, refining it to make it more sensitive to the needs of both.

The process of EIA should be treated analytically so that we can identify which areas are really of value. If this is done, as it must be, EIA will not decline, as Mr Mann's presentation might be taken to imply, but will be used increasingly in a direct and useful way, integrating the banking and engineering sectors, for example, with our very real and necessary environmental concerns.

F. H. Mann. EIA will certainly play a greater part in our lives in the future, but for a variety of reasons oil use will reduce and so too will its impact on our lives. This will be good to the extent we can replace it with gas and nuclear energy, but replacement with coal is likely to have adverse environmental effects.

D. W. Fischer. There are two dangers in particular in EIA. First is that individual scientists may take a proprietary interest in what they regard as being their piece of the environment, so that very quickly the holistic approach is lost. Second is the idea of searching for a perfect methodology, which is then of course based on perfect information, both of which do not exist. It is better if the decision process involves independent experts and does so openly, under powers more formal than in the UK. In the UK, the system is based on trust, which of course it is not in other countries.

O. C. Boyle (Department of the Environment, Dublin, Ireland). Professor Fischer analysed standard-setting from an organisational point of view. The UK approach is to set standards for the environment, then to use the environmental quality objectives approach towards emission standards or effluent standards, fixing the standard for each discharge so that it will match the capacity of the environment to assimilate pollution. In certain other countries a more rigorous approach is used where uniform emission standards are set regardless of the capacity of the environment. Has any research been done to determine which of the two approaches to standard-setting is most effective?

D. W. Fischer. I used the organisational approach because of the problem of standards not being scientific. Something very similar to the UK approach, of best practicable means (BPM), is found in Japan, Canada, the US, Norway and elsewhere. Where the UK does differ from other countries is in the local authorities having some final responsibility in the regard for local attitude about pollution inspectorate to adjust emissions to a local area. Otherwise there are more similarities than dissimilarities between different countries in the means for arriving at standards. In the UK, but more so in Norway, concern for the environment is new, so there is no historical research base.

C. S. Johnston (author). Professor Fischer concentrated on numerical emission standards, whereas the UK is thinking strongly of how to measure the quality of the environment. Local authorities try to protect the total environment. That means some research because the fauna and flora in a discharge site must first be ascertained. Thus one is trying to edge forward, legislating at the same time as gaining knowledge. This requires good communication and understanding between government and operator.

E. Marshall (Shetland Islands Council, UK). EIA and popular concern for the biological environment are in danger of getting out of balance. The UK is desperately short of engineers to gather the resources whose impact is of great concern. The education system produces many people to measure the impact of oil, but far too few to actually get it.

D. W. Fischer. The environment becomes the lion and eats the energy lambs! The shortage of engineers and other skilled manpower in Norway is probably the main factor limiting development of offshore oil resources.

A. D. Read (Petroleum Engineering Division, Department of Energy, London, UK). Offshore platform discharges may not provide the best example of standard-setting because there is little interest outside a small group of experts. Standards set for offshore platforms are essentially equipment performance standards. Although respected marine and fisheries scientists are consulted, they represent interest within government. There is inevitably a degree of arbitrariness on the initial setting of standards, which will then have to be revised in the light of experience. It is unlikely that extended consultation outside the present procedures will produce better or very different results.

D. W. Fischer. If independent experts outside government know that avenues are available to them to propound their views, they won't consider alternative avenues which may be extreme. If environmental scientists have a weakness then it is towards the environment. Indeed very often their personal and professional ethics become one. Sometimes rather extreme statements have been made. One way of avoiding such polarisation is to officially and formally give the individual access to the decision-making process, even if the information gained is marginal. Of course there has not been a lot of interest in the offshore field, but it is growing.

In the light of experience, is there now more research on alternatives, e.g. the production separators?

A. D. Read. There has been fairly encouraging feedback from the platform on how well the separators have worked. There is, however, an economic incentive to develop separators which take up less room on platforms. Progress is being made on designs which will achieve oily water throughput similar to that of existing units but in much less space.

H. Averley (W. J. Cairns and Partners, Edinburgh, UK). We have had an interesting description of two systems, but little on the difference between them in terms of relative scales of the two countries, the assumptions made by the style of government or the context in which decisions are taken. Scientists have said we have hardly got our foot on the first rung of the ladder of standard-setting.

The question of who must demonstrate that we are right to proceed with a certain energy or environmental conservation policy begs many questions about both the planning system and the environmental standards-setting system. Such general unease explains some of the concern in the UK that there is a conspiracy between government and the energy industry to pull the wool over the public's eyes.

D. W. Fischer. The doctrine of collective responsibility does not exist in Norway. Different government departments are seen to represent their own interests and not the collective interest of all of them as is the case in the UK. Task forces are used in Norway, and are closed systems, as in the UK, but the idea of the Norwegian being very independent and having a certain right to know, however ill-defined it may be, obliges the task forces to communicate their results to interested parties. Public information is also easier to obtain in Norway, because there is no Official Secrets Act. There are many other differences.

P. T. D. Patten. Perhaps one important difference is in the industrial history of the two countries.

Section III

PLANNING, LAND AND COMMUNITY

Chairmen

Papers 11 and 12
John Collins Esq., President, Royal Town Planning
Institute, UK

Papers 13 to 15
Professor J. Anderson, Department of Architecture,
University of Manitoba, Canada

Papers 16 to 18
Hugh Crawford Esq., Partner, Sir Frank Mears and
Partners, Edinburgh, UK

11

Europe and the Environmental Assessment

C. STUFFMANN

*Environment and Consumer Protection Service, Commission of the
European Communities, Brussels, Belgium*

ABSTRACT

*EEC environmental policy is based on a 1973 Action Programme (revised
1977) adopted because environmental resources are increasingly limiting
economic and social development. It also permits common objectives and
rules for environmental protection and improvement by Member States. A
central element of the policy is to prevent pollution or nuisance at source
rather than subsequently to mitigate undesirable effects, and it therefore
emphasises the importance of environmental planning. Member States'
governments, through the Action Programme, charged the Commission to
review and draw up proposals for harmonising national environmental regulations at the Community level.*

*Only France, Ireland and Luxembourg have specific environmental impact
assessment (EIA) legislation, of which the French, now in force for nearly
3 years, is the most systematic. Discussion is continuing with other Member
States, some of which already have rules for environmental controls and
extended public participation, but rules are often mutually inconsistent and
applied piecemeal.*

*The Commission therefore concluded that initiative at Community level
was needed and subsequently issued a 'proposal for a Council directive' which
stipulated that for important development projects in Member States an
assessment of likely significant environmental effects be completed. Before
an authority grants planning permission it must consult all others of environmental competence and organise public participation, including those in
neighboring states likely to be affected.*

The Commission feels all significant activities should be subject to EIA but the proposal is limited to individual industrial, agricultural and infrastructural projects because at this level divergent national rules may have a direct impact on investment decisions and, particularly, because such projects are in all Member States subject to certain authorisation procedures. The Commission was concerned to balance achievement of environmental objectives with flexibility of application without requiring complex changes in administrative structures and procedures.

The Commission thus achieved appropriate obligatory procedures for attaining environmental objectives, legal flexibility among the forms of Community legislation, and simplicity by not requiring new authorities or procedures.

Adoption of this directive in national legislation would improve decision-making for major development projects in the Community, and thus general investment climate, but not cause delay in planning or distortion stemming from differing national rules.

INTRODUCTION

The Commission of the European Communities considers that environmental impact assessment (EIA) can be one of the most important and efficient instruments for better and anticipatory environmental management. This is desperately needed, especially in Europe with its high density of population and industrial activity. A constant effort must therefore be undertaken to ensure widespread use of this instrument, the development and perfection of assessment methods, as well as the training of those who are to practise EIA.

For these reasons, EIA occupies an important place in the Community's environmental policy. In fact many people do not yet fully realise that the Community is not only a customs union with a common commercial policy and above all an agricultural policy but that it has started and is pursuing a very active environmental policy. Of course the European Treaties, conceived well before Rachel Carson's *Silent Spring*,[1] do not specifically mention environmental policy. But there are compelling reasons and arguments for a common action in that field.

First, the protection of the environment must be achieved at the most appropriate level and many problems can no longer be solved except in an international framework. The Community, in contrast to international organisations of the traditional intergovernmental style, has certain legislative powers which facilitate coherent and efficient action within a broader geographical context. Secondly, the existence of very different rules and objectives may well hamper the functioning of the Common Market and thus diminish the advantages of a large-scale economy precisely at a time when they are most needed. Thirdly, it becomes more and more evident that the resources of the environment are a factor limiting all further economic and social development. Environmental policy must then become an integrated part of economic policy. In our view, environmental preoccupations are therefore clearly an important element of the fundamental objectives of the Community, one of which is, according to Article 2 of the Treaty of Rome, harmonious economic development.

At the Paris summit meeting of 1972 the Heads of States and Governments called for environmental action at Community level. By November 1973, the first Environmental Action Programme of the Community was adopted and it was extended and continued in May 1977. Those programs defined a series of principles and objectives to be pursued at both Community and national levels, and defined actions for the Commission to elaborate proposals for submission to the Council for adoption. The main feature of the programs was the political commitment of Member States to deliberate and decide within the Council proposals submitted by the Commission. An impressive number of legislative actions have since been taken relating to water or air quality standards, product standards, e.g. for gas or noise emissions from motor vehicles, lead content in petrol, noise standards for aircraft, to regulations for the disposal of waste in general and toxic waste or used lubrication oil in particular, and even relating to the conservation of Europe's wild bird species. Almost all of these rules have been enacted in the form of directives, a legal instrument which defines principles, objectives and certain lines of application for the whole Community, but Member States, who must embody these lines in national legislation, have a large measure of freedom in implementation. It therefore combines in an ideal way indispensable harmonisation with necessary flexibility, allowing modulated application according to different national or regional structures and traditions.

EMERGENCE OF A PROPOSAL FOR ENVIRONMENTAL IMPACT LEGISLATION

One of the fundamental principles underlying the Community Action Programme is prevention. In 1973 the Council declared 'the best environmental policy consists in preventing the creation of pollution or nuisance at source, rather than subsequently trying to counteract their effects', and to this end 'effects in the environment should be taken into account at the earliest possible stage in the technical planning and decision-making processes'.

In 1977, aware of the development in the field of EIA in all industrial countries, the council stated that EIA was an appropriate instrument to meet the principles and objectives laid down in the Action Programme. It requested the Commission to examine to what extent harmonisation of relevant procedures in the Member States was necessary and whether such a procedure was needed at Community level and to make appropriate proposals.

All Member States already possessed long-established systems of environmental and development control and of land-use planning. In very different ways these made statutory provision for preparation of documentary evidence on certain environmental effects of proposed activities. Most of these provisions existed long before the National Environmental Policy Act was enacted in 1969 in the USA and the term 'environmental impact assessment' or 'statement' was introduced into the language of planners. The essential difference between those long-established practices and EIA is that the former normally concentrated only on certain aspects of the environmental impact of a planned project. They were more or less fragmentary in character, whereas the new formula means integrated and synthesised description and analysis of all environmental effects to be expected from the project under scrutiny.

EXISTING ENVIRONMENTAL LEGISLATION

Although existing procedures in all Member States have been strengthened to take environmental consequences into account, only three countries have specific legal provision for EIA: France, Ireland and Luxembourg. Even those provisions, however, differ widely between countries and do not cover all relevant elements.

France has the most systematic and complete legislation for individual projects. The principle of impact studies was introduced in July 1976 by the Nature Protection Act, Article 2 of which stipulates that 'studies undertaken prior to commencing significant public works or private projects requiring public authorisation must include an impact assessment study'. Legislation in force since 1 January 1978 specifies methods of application including technical and financial thresholds. For minor projects, a single impact notice is required. More than 4000 impact studies have been prepared so far. A seminar in Paris last June gave an opportunity to discuss experience gained since enactment of the legislation. The general conclusion was that the running-in period had been satisfactory and convinced authorities as well as project promoters of the advantages of legislation.

In Luxembourg the Natural Environment Law 1978 and the law relating to Industrial Establishments stipulate, at the discretion of the competent Minister, that impact assessment studies should be prepared for industrial or infrastructural projects which may have significant effect on the natural environment. In contrast to the French system the law in the Grand Duchy does not include effects on the urban population.

In Ireland the Local Government Planning and Development Act 1976 provides for preparation of a written environmental impact study as part of the development control procedure for large private projects which are expected to cost at least £5 million. However, the authorities have discretion in determining whether to insist on EIA. Public works are not submitted to specific mandatory impact studies.

In the Federal Republic of Germany there is a complex and elaborate system for authorisation of major activities likely to have environmental impact. However, the provisions referring to impact assessment are not systematically grouped. In 1975 the Federal government issued a Cabinet Order obliging all Federal administrations to prepare impact studies for all projects within their competence. This decision, however, is not binding on the Lander (provinces). Only the Senate of Berlin has introduced a similar obligation to its own administration. The efficiency of this procedure is uncertain as application of it depends on internal administrative support. In addition there is no specific requirement for publication let alone public examination of the documentation.

The Netherlands will probably be the next Member State to adopt EIA legislation. In 1976 the Dutch Study Council on the Environment recommended to the Dutch government that EIA should be compulsory for a wide range of governmental measures and activities. In response, the Dutch government instituted a comprehensive program of case studies to test the practicability of these recommendations. In 1979, after successful completion of the case studies in which the European Commission participated in a modest way, the government indicated the main lines along which in intended to prepare draft legislation for submission to parliament. Though it is intended to restrict the number of projects requiring full EIA, the law will probably include land-use planning within its scope.

In Belgium and Italy environmental impact is partially taken into account within the existing provisions for authorisation of projects. In Belgium the government has announced its intention to introduce a draft law on EIA in 1980.

In Denmark provisions concerning environmental control and land-use planning were completed in 1970 in order to take better account of environmental aims. They now provide for certain of the requirements of comprehensive EIA but there is no general obligation for preparation of environmental impact studies.

In the United Kingdom there has been undoubtedly major interest in environmental questions for a number of years, which has materialised in a series of most important studies, e.g. those by Dobry[2,3] and Catlow & Thirlwall,[4] the University of Aberdeen and the Royal Commission on Environmental Pollution. Strict procedures are laid down in the Town and Country Planning Act 1971 and the widespread implementation of this legislation in public inquiries is appreciated.

Although environmental impact studies have very usefully been undertaken in the UK, the attitude of the government still seems cautious, favoring a pragmatic rather than a mandatory approach, with a view to restricting the use of EIA to very few projects. There is of course much to be said in favor of that approach as far as it is necessary to save resources and to debureaucratise society. On the other hand, pragmatism very often leads to inconsistency and the mandatory preparation of these studies is intended to ensure that all development projects are systematically checked for their overall environmental consequences.

Initiatives taken at Community level must be understood against this background of national practices. The Council of Ministers has recognised that EIA is a means of achieving the principles of a common environmental policy. The incomplete cover and content of present practice in the Member States may well justify an initiative at Community level.

Moreover, and this is important, there is a direct link with the correct functioning of the Common Market, which is a more fundamental responsibility of the Community and the Commission. Significant variations in the extent to which environmental preoccupations are taken into account in economic decision-making could, in the short-term, favor investments in regions or sectors where such procedures were most permissive. Too great a divergency in enforcement of environmental standards may therefore well lead to some distortion of investment decisions and hence of competition within the Community. On the basis of these considerations, confirmed by the studies referred to above, the Commission decided that execution of the Action Programme called for Community initiative. After many years of deliberation and consultation with governmental experts, non-governmental organisations from industry as well as conservation scientists and others, the Commission adopted and tansmitted to the Council the proposal for a Council directive concerning the assessment of the environmental effects of certain public and private projects. This proposal was intended as a first but substantial step towards establishment throughout the Community of a framework of minimum requirements for assessing industrial, agricultural and infrastructural projects likely to produce significant effects on the environment. The main feature is an attempt to strike an appropriate balance between consistency with the Community's objectives and flexibility, allowing a modulated implementation to take into account differing national and regional administrative and economic structures.

Another feature nearly as important is that the directive should not lead to creation of new authorities or procedures. The proposal was therefore limited to the introduction of certain common principles into procedures already existing in all our Member States. All provisions in the proposal are dictated by these three main elements: consistency, flexibility and administrative simplicity.

PRINCIPAL PROVISIONS OF THE PROPOSAL

The Area of Application

The proposal is confined to authorisation of individual projects although in the Commission's opinion

> 'in a coherent system, provisions for the advance assessment of environmental effects should obviously be present at all the administrative levels at which these activities are controlled, given the inter-relationships existing between them; at the level of authorisation procedures for projects, for the preparation of regional programs, land-use or economic plans and the procedures for licensing certain products'.

The Commission's decision to limit the proposal to individual projects is mainly for practicability, as the existence of authorisation procedures for individual projects in Member States contrasts with the rudimentary state of plan assessment methods and incomplete and widely differing coverage of land-use and other planning systems in the Member States. Moreover, it is certainly on the level of individual projects that inconsistency of national procedures may directly affect investment planning and through this the normal functioning of the Common Market. Action in that sphere is therefore to be considered an elementary Community responsibility. However, the Commission simultaneously confirmed the general orientation for the wider application of EIA.

As to individual projects, the proposal obliges Member States to ensure that, before any planning permission is given, projects likely to have significant effects on the environment be subject to an appropriate impact assessment. These are listed in Annex I of the proposal. The competent authority for determining whether a project needs such an assessment depends on the type of project and Member States are completely free in the designation of these authorities.

However, the proposal stipulates that for all categories of projects mentioned in Annex I, the full assessment should be undertaken, except for some marginal cases which may be exempted with the agreement of the Commission. The full assessment would also be compulsory for categories of projects listed in Annex II, which lists projects whose impact would be less serious than those listed in Annex I, as well as modification of projects listed in Annex I, unless they are exempted in accordance with thresholds or criteria to be defined by Member States. These thresholds or criteria would have to be notified to the Commission to enable them to take further steps in case of excessive disparities between practices in the different Member States or between practice and the fundamental principles. An EIA would equally be required for those projects outside the categories mentioned in Annexes I and II if they might affect zones of particular environmental sensitivity.

Preparation of the Environmental Impact Assessment

The proposal stipulates that the developer has the primary obligation to supply, simultaneously with his application for project authorisation to the competent authority, all relevant information required in an appropriate environmental impact study. Annex III details the information to be included:

(i) description of the proposed project and reasonable alternatives to it;

(ii) description of the environment likely to be significantly affected by the project;

(iii) assessment of its likely significant effects on the environment;

(iv) description of any environmentally mitigating measures that are proposed;

(v) indication of the likely compliance with existing environmental and land-use plans and standards for the area;

(vi) justification, in the case of any project likely to have significant adverse effects on the environment, of the rejection of reasonable alternatives expected to be less environmentally damaging;

(vii) a non-technical summary.

The reason for placing this obligation on the developer is that he is the one who knows the project but it will also alert him at the earliest stage of the planning process to account for environmental consequences and so lead him to search for the best solution before submitting his application for authorisation. Of course the competent authority will have to assist the developer in the preparation of the study and check that it is sufficiently complete with reference to the data normally available to the developer or try to make more complete information available if the need arises. The information required may also relate to the impact on the environment of another Member State.

The proposal requires, as a second phase of the assessment operation, that the competent authority for authorisation must consult all relevant administrative authorities and other statutory authorities or bodies with specific responsibilities for environmental matters. Other Member States whose environments may be significantly affected by the project must also be consulted. Moreover, the competent authority must publish the fact that the application for authorisation has been made, make all the documentation available to the public and make appropriate arrangements enabling the public to express its opinion on the project.

Final Assessment

When taking its final decision on the proposed project, the competent authority will have to make its own final assessment based on the information supplied by the promoter and that gathered through consultation with other authorities and the public. This final assessment is to be published unless planning permission is refused for other than environmental reasons. This publication must contain the assessment study, a synthesis of the essential comments put forward during the consultation processes, and any conditions attached to the planning permission. Unlike the EIA procedures in the USA, the proposed procedure does not provide for a second public consultation on the final decision.

Monitoring

In the case of a proposal being authorised, the competent authority is bound to ascertain periodically whether the conditions attached to the planning permission are being fully respected and also whether the project is having environmental effects which had not been dealt with in the assessment process. This monitoring should contribute to evaluation of the quality of past impact studies and thus to the improvement of the assessment methods.

The Commission also intends to make a contribution of its own to further methodological progress by conducting a series of pilot studies for certain types of project, which then will be published. In addition, the Commission is presently drawing up a pilot method of ecological mapping destined to permit easy access to all relevant information on environmental characteristics and above all on the ecological potential of a region. Such a mapping should evolve into a most appropriate instrument for environment-related development planning and should supply planners with a practicable and sound basis for all EIA studies. This new method, which has been designed on the basis of experience in different nations, has been tested in ten case studies and we hope next year to be able to submit to the council proposals for making use of this planning tool.

CONCLUSION

In summary the Commission has, in accordance with the mandate given by the Community Action Programme, concluded that action to bring about coherence in the use of EIA should be taken at Community level for better functioning of the Common Market as well as implementation of the fundamental principles laid down in the Action Programme. In preparing its proposal it has constantly tried to strike the most appropriate balance between consistency with these two major preoccupations, and maximum flexibility and administrative simplicity.

Consistency. Practically all elements normally considered essential to an appropriate impact assessment will be included in national planning procedures. That should enable authorities to make decisions in the full knowledge of overall environmental requirements and result in sufficiently equal treatment of investors in all Member States.

Flexibility. The legislative form of the directive ensures maximum flexibility by defining only principles and broad lines of action, leaving method of application to the national authorities who are thus able to account for the situation in their separate countries.

Administrative simplicity. Apart from the restricted field of application, the Commission's proposition carefully avoids all that would imply creation of new authorities or procedures. Its scope is limited to the introduction of a series of common principles to planning permission procedures which already exist in the Member States.

As these national procedures have historical roots and so many are rather complex, the insertion of the Community principles may present, as experience suggests, opportunity to review, simplify and rationalise old administrative procedures. There are also many practical examples that show clearly that appropriate undertaking of EIA improves the decision-making process and contributes to shorter procedural delays, rather than the reverse, and facilitates the consensus which is nowadays needed for the realisation of major projects. The eminent legal experts whom we have consulted were of the opinion that the application of our proposals should lead to less litigation than at present.

The cost which will be incurred by carrying out the proposal should always, according to experience, be within 1% of the total investment, which seems fully acceptable given the advantage of improved planning and decision-making not only to the Community in the form of reduction of social costs arising from imperfect planning but also to the investor in the form of smoother planning procedures and more appropriate design of the project from the outset. There are plenty of examples to show that a good EIA has allowed the project promoter to considerably reduce the cost of his investment compared to initial plans.

What is happening in Scotland may be the best practical example to bring home the importance of this instrument, if we really want to reconcile economic development with conservation of the environment. They are inseparably connected and both are equally elements of quality of life.

REFERENCES

1. Carson, R. (1962). *Silent spring*. Hamish Hamilton, London, UK.
2. Dobry, G. (1974). *Review of the development control system. Interim report.* HMSO, London, UK.
3. Dobry, G. (1975). *Review of the development control system. Final report.* HMSO, London, UK.
4. Catlow, J. & Thirlwall, C. G. (1976). *Environmental impact analysis.* Research Report 11. Department of the Environment, London, UK.

12

Ecological Planning

I. L. McHARG

*Department of Landscape Architecture and Regional Planning,
University of Pennsylvania, Philadelphia, USA*

ABSTRACT

*Ecological planning is the application of physical and biological science
within an ecological model. An integrated understanding of the environment
is achieved through geology, meteorology, soil science, plant and animal
ecology. Planning involves use of this model to find the optimum environment
for people and their activities. Failure of any biological system, including
human activity, to find the environment to which it is best fitted ultimately
leads, in Darwinian terms, to competitive exclusion.*

INTRODUCTION

Ecological planning, unlike statutory planning, is simply physical and
biological science within an ecological model. It says that it would be
nice to know what you are doing before you do it. This means the operation of the biophysical world must be understood in order to predict the
consequences of contemplated actions.

Statutory planning seems to be administration, in which there is
believed to be wisdom in a plan, and the planner's function is to admi-
nister. That is not planning at all. In addition, planning as practised in
the United States and more so in Britain has not been touched by biophysical science.

THEORY OF ECOLOGICAL PLANNING

The theory of ecological planning revolves round three linked terms, syntropy, fitness and health, the antitheses of which are entropy, misfitting and morbidity.

Syntropy and entropy are generally well understood. Every energetic transaction increases entropy. In any energetic transaction, including the activities of all life forms, the entropy of the system is increased. But there are some energetic transactions as a result of which method and energy in part of the system is increased.

All life's processes are energetic but none the less they proceed up the phylogenetic scale. Each step, through algae, liverworts, mosses, ferns, etc., has resulted from an increase in entropy yet each step represents more highly ordered matter in organisms. Creativity, or syntropy, then is really a description of evolution, which can thus be measured in terms of entropy.

Fitness has two meanings, derived from different but complimentary sources. The first is due to Charles Darwin who stated simply that the surviving organism is fit for the environment. Lawrence Henderson introduced a complementary concept, that there is an infinite variety of environments and the opportunities afforded by them are complementary to the capability of the organism fitting an environment. Both statements describe an evolutionary imperative for every biological system, whether a cell, organ, organism or ecosystem, to find the environment to which it is best suited and to adapt itself to it. The meaning of fitness is, of course, implicit in the definition itself. The fitness of an environment is defined by the degree to which the environment, as found, is doing the largest part of the work required by the consumer. The consumer has then to import the least amount of energy to make its environment more fit. There is clearly a thermodynamic imperative in evolution which requires all systems to find the fittest environment and to adapt them continuously. It is a creative process and the amount of energy employed is a measure of the fitness of the environment. Of two complementary systems, the one which uses less energy to achieve a fit environment will undoubtedly, in Darwinian terms, survive. The prodigal system, like ours in the West, is likely to succumb.

The concept of health is implicit in Darwin's proposition of fitness. If a system has little morbidity it has probably been able to find the fittest environment and adapt it to its own requirements; this capability of creative, syntropic fitness is revealed by health. Any system displaying signs of morbidity is probably incapable, permanently or temporarily, of finding a fit environment or adapting its environment to suit its own needs.

ADAPTATION

The stark simplicity of the above model is beautiful. The model clearly revolves around adaptation, and the three instruments of adaptation must be understood.

The first instrument is a physiological adaptation, through mutation and natural selection. People have little option they can exercise in this respect beyond spouse selection, and generally rationality is little used for that. The second instrument is behavioural adaptation. Drugs provide one means of doing this. However, few people, if any, seriously believe that we should try to modify inate behaviour.

This leaves the third instrument for successful adaptation, cultural adaptation, for which the means most directly available and corresponding to the Darwin–Henderson evolutionary interpretation is planning. Obviously the most important quest for all men and all institutions is survival, for which creative fitting is a condition. The instrument for achieving it is in fact planning, so clearly planning is the single most important human activity.

BIOPHYSICAL ANALYSIS

Too often planning has been done with no understanding whatsoever of our biophysical environment. Thirty years ago, and to too great an extent today, geology, geomorphology and ecology were unknown to planners, whose main concern was visual analysis.

Any biophysical analysis for planning must start with an understanding of geological phenomena, their history, processes and dynamics. The meteorological characteristics of the region in question must also be known, both historically and in modern terms so that, for example, the dynamics of climate associated with acid rain are known; an understanding of bedrock geology and so on through ground water hydrology, surface water hydrology, soil science, plant ecology and animal ecology must be gained. This assembly of information really uses chronology to explain reality. With the information from the various disciplines can be built an interacting biophysical model, with as much predictive capability as possible.

The next step in the planning process is to ask of the consumer which aspects of the environment are propitious and which are detrimental. Then with the aid of a computer, or by overlay mapping, it is a conceptually simple task to ask the model where all or most of the propitious factors exist and none or few of the detrimental ones do. Darwin's evolutionary imperative defines the fittest environment as that where least adaptation is required of the environment or of the incoming system in order for the latter to survive. The model solves a very substantial part of the problem.

SOCIAL ANALYSIS

It is paradoxical that in science it is widely believed that nature is systematic. Biophysical realities are understood to be a result of evolution and the operation of natural laws and time; but it is not believed that man is systematic and still less that there is any systematic relationship between man and nature. Yet in reality the contrary is true. For planners, it is necessary not only to build the biophysical model but to populate it with people in a causal way. Planners must understand, however, that people are where they are, doing what they are doing, for some very good reason. Planners must therefore know what people's needs and desires are and relate them to the environment in general. They must find which of all available environments is the most fit for their needs and help them accomplish the process of successful adaptation.

Understanding the reasons for the distribution of people and their activities requires an anthropological model. This model presumes that a natural system is a system of social values. It is also a system of resources, but what constitutes a resource varies with the perception and technology of the viewer. Nevertheless, given perception and identification of some resource, and technology to use it, then the resource locates the means of production.

The long social history of the United Kingdom may obscure these fundamental relationships and the United States perhaps provides a better illustration. In years gone by the means of production could not generally be satisfied by available articifers, so it was necessary to send abroad for them. Such people had not only an occupational identity, but often also ethnic and religious identities. So the first people to mine coal in Pennsylvania were Scotsmen, probably selected because of their experience as miners. With them they brought their characteristic institutions, settlement patterns and values. This meeting of particular people, culture and environment was not entirely accidental, and is clearly evident in Pennsylvania today.

Given the detailed knowledge embodied in the models, we are able to predict the likely consequence of any contemplated action. Understanding ethnographic history also permits an understanding of power.

So planning should be based on a model which integrates the disciplines of geology, meteorology, geomorphology, pedology, plant and animal ecology, anthropology and sociology. Perhaps the best way of testing the efficacy of ecological planning, however, is to undertake an epidemiological study. If people are healthier as individuals (or as families, institutions, etc.) than their competitors, then planning has been successful.

13

St Fergus: Visual and Design Considerations†

F. E. DEAN

British Gas Corporation, London, UK

ABSTRACT

British Gas has expended considerable effort over the past decade in developing and using environmental impact assessment (EIA). EIA was used in selection of possible sites as landfalls for gas from the Northern Basin of the North Sea. A proposed terminal at Strathbeg/Crimond was rejected because, although the site had ideal visual, technical and engineering characteristics, it would have been in conflict with an internationally important nature reserve and a Ministry of Defence radio station. The site finally selected nearby, St Fergus, did not present such conflicts. However, special precautions were necessary to limit noise, obtain adequate water supply, deal with site drainage and reinstate coastal sand dune systems. Adverse effects of a large construction labor force on the local community had to be minimised. Proposed extensions to the existing site have also been subject to full EIA.

PLANNING BACKGROUND

The national and international need for energy is often in conflict with the national desire for conservation of the environment, the proposals of

† This paper was presented with Mr G. Graham CBE Architects Design Group, Lockington Hall, Lockington, Derbyshire DE7 2RH, UK; Past President, Royal Institute of British Architects whose contribution to the manuscript was not available for publication.

national and local authorities and the substantial needs for agricultural production. In an effort to identify major technical, economic, social, ecological and amenity conflicts and to propose suitable solutions to them, British Gas has devoted much time and effort in the last 10 years or so. This type of work is now recognised throughout the world under the name of environmental impact analysis (EIA). The Corporation may feel some sympathy with the man in the Molière play, who suddenly became aware of the fact that he had been writing prose for 50 years. The traditional name of Site Location and Assessment Procedures may be more appropriate and less likely to be accused of being quasi-scientific.

There are probably as many definitions of EIA as there are planning authorities, major developers and consultants, and all are different. At the request of the government of the day, Mr George Dobry, QC, made an investigation of the subject as a factor of development control in the planning system. In his report[1,2] he gives what is probably one of the best:

'the orderly assessment of the overall impact of a planned or existing development on the environment in terms of both physical and socio-economic effects'.

The general location of a gas industry development is determined by

(i) The geographical location and size of the offshore fields;
(ii) the technical requirements of the national transmission system which in turn is shaped by the requirements of British Gas's 14 million customers.

These parameters, however, give only a broad indication of areas in which suitable pipeline routes and sites for above ground installations have to be found. The exact location, which must be suitable from an engineering point of view, is often determined by environmental considerations which cannot be defined until detailed studies have been carried out in a wide range of subjects ranging from ornithology and geomorphology to architecture and socio-economics. Although not all of these subjects may be appropriate in every case, the questions have to be asked and the answers carefully considered because values can change from site to site.

It might well be thought that all this is slightly academic and that the Corporation could well pay little attention to these matters and rely simply on a case based on the national need for additional energy supplies. Unfortunately this is not so. Governments of both political persuasions have made clear their desire to ensure that environmental considerations are not subjugated unnecessarily to short-term economic benefits. Studies of the EIA concept were initiated by the government and reports issued.[3] The main conclusions were that EIA was of value for certain major projects and a list of some 30 types of development which should be subject to this form of examination was issued. Included in this were coastal reception terminals and gas compressor stations. The EEC has at last issued a draft directive (Stuffmann, these proceedings) which would make the use of EIA a mandatory part of the planning procedures of the Member States. British Gas, whilst acknowledging that EIA has a significant and valuable role to play, feels that the permissive method used in the UK meets all requirements very satisfactorily.

It is against this background that the development of the Northern Basin gasfields should be considered.

THE ST FERGUS PROJECT

It became clear in late 1969 and early 1970 that there was every probability of gas being discovered in commercial quantities in the Northern Basin of the North Sea, and that, if it were, it would come ashore in the north-east of Scotland. Consequently, on a contingency planning basis, a preliminary search was commenced along the northern and north-east coast of Scotland as a continuation of the work already done in north-east England. This study revealed a number of potential sites including those now known as Crimond and St Fergus.

In 1972 the Frigg field was announced as being of commercial size and the decision to develop it as rapidly as possible was taken. There were two possible landfalls for the undersea pipelines, both of approximately equal length, one in the Wick/Thurso area with a potential site for the terminal at Sinclairs Bay and the other in the Fraserburgh/Peterhead area. The extra onshore pipeline length rapidly ruled out the former and a study of the immediate offshore seabed conditions was commenced southwards from Fraserburgh towards Peterhead. This survey soon revealed a good pipeline landfall in the vicinity of the Loch of Strathbeg, on the landward side of which was the disused airfield of Crimond which would provide a suitable site for the necessary terminal facilities.

The Loch of Strathbeg/Crimond Proposal

Geographical considerations. Geographically this site had much to commend it. The sealine would be probably the shortest possible and the

sandy beach and immediate offshore seabed provided excellent con-
conditions for pipe laying and burying. Adjoining the beach is an extensive and unstable sand dune system, the dunes and the adjacent Loch of
Strathbeg having been formed comparatively recently in 1720 during an
exceptional storm which blocked the outlet of the local burns. The loch,
though extensive, is 4–5 ft deep and bounded on the inland side by a large
area of well drained flatland at a slightly lower level than the surrounding
terrain.

Technical considerations. Technically, the site was probably ideal. With
the shortest sealine and good landfall conditions, economics were also
in its favor. Admittedly, the crossing of the unstable dune system might
cause problems but these could easily be solved by stabilising them. The
shallowness of the loch itself meant that it would provide no obstacle to
the pipeline route to the airfield which, viewed from the civil engineering
aspect, would be suitable for industrial development. However, there was
only scope for one offshore pipeline.

One major obstacle was foreseen and that was the Ministry of Defence
interest in the site which they wished to use for a strategic communications transmitting station. Discussions were held with them to ascertain
if it would be possible to use adjoining portions of the site or even a joint
use, and at one stage it seemed that a compromise might just be possible.

Environmental considerations. If technical considerations indicated
that this was the optimum site, environmental factors led to the exactly
opposite conclusion and a tremendous opposition built up to its use as a
terminal. There had been similar antagonism to its use as a radio station
but this battle had been lost and planning consent granted.

Having said that, and before dealing with the undoubted environmental
disadvantages, it must be emphasised that there would be certain environmental benefits arising from the use of the airfield site. First, the land was
only used for low grade agricultural purposes, the runways for motor racing
and some form of industrial development had already been proposed.
Secondly, it was almost certainly the only site out of several alternatives
which were examined which could be almost completely screened from
areas to which the public had reasonable access. There is in existence a
very good belt of trees between the site and the main road and this is a
major factor in an area not well endowed with woodland and where the
rate of tree growth is very slow indeed.

For the numerous, well informed, articulate and influential objectors the main platform was the status of the Loch of Strathbeg, a Grade 1 site of Special Scientific Interest, and the associated dune systems, a Grade I coastal site. Together, these are of ecological importance on three separate counts:

(i) the ornithological importance of the Loch as a staging post for many thousands of migratory birds such as Pink-footed Geese and many other species;

(ii) the geomorphological nature of the unstable dune system, one of a very few in western Europe;

(iii) limnologically, the loch is unique because it is brackish and shallow. These characteristics have produced a special ecosystem.

These factors produced an immediate and predictable reaction from local and national conservation interests which was centered on and coordinated by a group known as the Environmental Liaison Group, the secretariat of which was provided by the biology and geography departments of Aberdeen University. This group was recognised by the then Aberdeen County Council who used it as a consultative and advisory committee. Apart from this semi-official body there was also considerable local feeling against the proposals, which was made clear in no uncertain terms at the local meetings which were held in the villages of Crimond and St Fergus.

As a result of this attitude, very extensive consultations were held not only at local level but at regional and national levels. In fact, these negotiations represented the start of the present regular and amicable consultation procedure which has been established between British Gas, the Scottish Development Department and the Regional and District Council. These involve discussions on matters of common concern not only relating to immediate planning problems but also to long-term contingency planning.

The decision. All these discussions and negotiations occupied a period of some months but it ultimately became clear that, quite apart from the ecological damage that might or might not have been inflicted on the nature reserve, the Ministry of Defence were not prepared to move from the site which they owned and for which they had a valid planning consent. The decision was therefore made to re-open the search for an alternative landfall, which revealed that there was in fact a short length of coastline just south of Rattray Head which was capable of providing suitable pipe laying conditions for several pipelines. This landfall, near the village of St Fergus, was adjacent to a potential terminal site which had been identified in 1970.

After taking all these factors into account it was agreed that, in view of the implacable stance taken up by the Ministry of Defence, which would cause lengthy delays to the project, the Crimond site should be abandoned in favor of that at St Fergus. Whilst the ecological problems would have been extremely difficult and expensive to solve, it is still felt that solutions acceptable to all parties could have been implemented if the alternative decision had been taken.

The St Fergus Site

Geological and technical considerations. Geographically, the site is similar to that at Crimond; the sealines are marginally longer but on the other hand the landlines are correspondingly shorter. The dune and slack system is only superficially similar to that at Strathbeg, but is essentially stable and therefore of a lower order of ecological interest. Agriculturally the land was of a slightly higher quality but still largely used for cattle grazing although some arable crops were grown. The soil conditions were not as good as expected and necessitated the removal of thousands of cubic yards of peaty material. The transport and disposal of the spoil created some environmental and social problems for which solutions had to be found.

The general technical development of the site has been described in detail elsewhere[4] so in this paper comment will be restricted to those factors with an environmental content, i.e. noise, drainage, water supply, the provision of a labor camp, and the possible interaction with the radio station at Crimond.

It was realised from the outset that noise could be a major problem with an installation containing one of the largest compressor stations in the world as only a relatively small part of the whole complex. The ambient noise level survey showed that under still atmospheric conditions, which do not occur very often in this area, the existing noise levels can be very low indeed. Lengthy and detailed discussions were held with the environmental health officers of the local authorities and ultimately noise levels were agreed for the site boundaries of approximately 50 dBA. In order to achieve these levels great care had to be taken in the design of the plant:

   (i) to suppress noise at source whenever possible, e.g. modified fan design for the after-coolers;

  (ii) to place potentially noisy items of plant in the more remote parts of the site when practicable;

 (iii) to install adequate silencing where necessary.

Decisions on these three items are not easy to take and require very careful assessments of design, economics and timing.

It was recognised that site drainage would require special treatment. The site was crossed by several burns which discharged ultimately into the Annachie lagoon. It was decided that these should be amalgamated into a common flume which traverses the site. The surface water flowing through the flume passes through oil separators for pollution control (and the fire ponds to keep them full) before finally leaving the site, but the level of the tide still controls the rate of water clearance under all weather conditions.

An adequate supply of mains water to the site has been a headache from the outset. The whole of north-east Buchan has suffered from an inadequate water supply for some time and in the initial construction period strict limits had to be applied. At certain times this meant a restriction on such operations as concrete mixing.

The combined population of St Fergus, Crimond and Kirktown is 400–500. The introduction of a labor force which would exceed this by a factor of two or three gave rise to concern. It was agreed that the only possible solution would be to provide a semi-permanent camp which would be completely self-contained and provide within itself for all the requirements of its inhabitants. This camp, which is still in occupation some 6 years after its construction and likely to be for some years yet, was established on part of the airfield at Crimond. A feature of such camps is often a conflict of interest with local residents and authorities. To avoid such conflict a joint coordinating committee was set up from the outset with representatives of the operators, contractors and local interest including the police. As a result of the existence of this committee there have been no significant incidents during the life of the camp. A similar committee was also set up to deal with the environmental and amenity matters relating to the construction and operation of the terminals and this also has been successful.

At the time of the original planning applications for the St Fergus site (1972) it was realised that its presence might have some interaction with the operations of the Crimond radio station. Discussions on this point

were held with the Ministry of Defence when it was agreed that provided
the gas installations did not exceed certain height/distance parameters no
difficulties would arise. Five years later when both installations had been
commissioned it was suggested that there was a possibility that electric
currents might be induced in the equipment at St Fergus by the radio
transmission from Crimond. The supposition was taken very seriously
for obvious reasons and an exhaustive investigation was carried out by
all interested parties including the Health and Safety Executive. The outcome was successful in as much as it was demonstrated that although it
was theoretically possible for such currents to be induced, in practical
terms their possibility was remote and their magnitude so small that no
hazard could exist.

Ecological considerations. Ecologically, the interest of the site was minimal apart from the dune system, in particular the dune slack area. The
planning authority was insistent that disturbances to them be minimised
and that effective reinstatement was of prime importance. The cuts
through the dune, one for Total's two lines and one for that of Shell, were
done most carefully and the sand replaced almost grain by grain. Bitumen
spraying was used successfully to prevent wind blow, which can be very
damaging to dune systems. The replanting was done by hand under the
supervision of, with labor supplied by, the Botany Department of
Aberdeen University. Another planning condition was that the whole
dune system should be managed as a conservation area and to this end a
working party has been set up under the chairmanship of British Gas with
representation from Shell and Total, Grampian Region, Banff Buchan
District Council, Aberdeen University, Nature Conservancy Council,
Royal Society for the Protection of Birds and the Scottish Wildlife Trust.
It is hoped that the reserve will be in full operation within the next year
or two.

Future development. It has long been British Gas policy that the
number of coastal terminals should be kept to the minimum and to this
end it was always envisaged that St Fergus could be developed beyond
the requirement for Frigg gas. In fact, before this phase had been
completed work had started on the installations required to handle the
associated gases from the Brent field. The gas gathering pipeline system
study recently submitted to the government recommends the construction of a new reception terminal immediately to the north of the existing
site. A planning application for this was submitted in May 1980. As sup-

porting documentation a full EIA was submitted with an in-depth visual impact analysis. This application also includes provison for a substitute natural gas plant on the same site using natural gas liquids as feedstock.

REFERENCES

1. Dobry, G. (1974). *Review of the development control system. Interim report.* HMSO, London, UK.
2. Dobry, G. (1975). *Review of the development control system. Final report.* HMSO, London, UK.
3. Catlow, J. & Thirlwall, C. G. (1976). *Environmental impact analysis.* Research Report 11. Department of the Environment, London, UK.
4. McHugh, J. & Marris, J. H. S. (1976). *The engineering of the Frigg project.* Communication 990. Institution of Gas Engineers, London, UK.

14

Flotta Terminal: Visual and Design Considerations

M. SARGENT

Orkney Islands Council, Kirkwall, UK

ABSTRACT

In development of the oil handling terminal at Flotta, Orkney, a staged environmental impact assessment (EIA) was integrated with the design of the terminal. The procedure provided a focal point for close cooperation between the local authority, developer and consultants. The EIA framework enabled sophisticated design techniques to be applied to the major problem of visual integration, with maximum effect. The result was rapid statutory approval, implementation and successful integration of a major industrial site into the beautiful but sensitive Orkney landscape.

INTRODUCTION

The discovery in 1973 of major oil deposits in the Piper field of the North sea, some 100 miles east of the Orkney Islands, triggered a search by the Occidental North Sea Group for a suitable oil terminal site. The apparent advantages offered by locations in the Orkney Islands required rapid consideration by the local planning authority of immediate and long-term implications of any major industrial complex, for the community and the environment.

The UK has a statutory system of land-use planning, by which development, or change of use of land, requires, with certain exceptions, planning permission. Although every application is assessed on its individual merits, there is a system of development plans which set out land-use zones. In the UK, North Sea oil development in the 1970s coincided with revision of the planning system.

In 1973 the existing county development plan for Orkney needed revision to accommodate the possible ramifications of oil development. The Orkney County Council under the new structure plan proceedures prepared an interim strategy plan.[1] The plan identified likely types of oil development, their acceptability to Orkney, those areas of the Islands requiring protection of special interests and those areas with development potential, incorporated them into policies for environmental protection and improvement, and later into the Orkney structure plan.[2]

The Island of Flotta, Scapa Flow, with year round deepwater access' and sufficient land to meet long- and short-term requirements was identified by the Council's consultants as a potential industrial site.

ENVIRONMENT ASSESSMENT PROCEDURE

Occidental's search for a suitable site for an oil handling terminal identified nine possibilities in north Scotland, including the Orkney Islands and Shetland. Feasibility studies short-listed three sites from which the discussions with the Orkney Islands Council, the Development Department and the Highlands and Islands Development Board selected Flotta as acceptable for a terminal.

It was critical for Occidental to combine fast development of the terminal with a proper regard for the social and natural environment. Both the planning authority and Occidental considered it essential that recommendations for environmental protection be incorporated into design of the terminal.

The environmental consultants (W. J. Cairns and Partners, Edinburgh) were engaged by Occidental to undertake an environmental appraisal and to work in conjunction with the project engineers and mangers. A preventive strategy was adopted based upon *identification* and *evaluation* at the outset of potentially significant environmental impacts, *protection* of the environment by all practicable measures, and *resolution* of conflicts between social, visual, ecological and engineering requirements. A staged environmental impact assessment (EIA) was therefore integrated with the design process. Five stages were undertaken:

(i) project proposal—a statement of intent;
(ii) environmental assessment—pre-development survey and identification of potential impacts;

(iii) visual impact analysis and landscape proposals (detailed assessment and design studies);

(iv) marine ecosystem impact and protection proposals (detailed assessment and design studies);

(v) terminal operation manual.

Following initial consultation with the technical officers of the local planning authority and the advisory agencies, documents prepared as part of this procedure were formally submitted to the planning authority in support of the statutory planning requirements. The preparatory informal contacts contributed critically to close cooperation between the Orkney Islands Council and Occidental, through their environmental consultants, and the rapid processing of the application.

The Flotta initiative has emphasised the increasing general recognition of the substantial economic benefit of rapid approval and implementation of major capital projects. The scale of investment involved and the national importance of the Flotta projects were significant.

PLANNING PROCEDURES

Application for outline planning consent for an oil handling terminal on the Golta Peninsula of Flotta was submitted in July 1973, supported by the Stage I report of project proposals.[3] The environmental consultants' initial findings, Stage II report of environmental assessment, provided further supporting information.[4]

Statutory planning procedures required advertisement of the application, which resulted in two formal objections from environmental groups. The objections were withdrawn following consultations, and six technical representations were also satisfactorily answered. The application required amendment to land-use zones authorised by the Secretary of State for Scotland in January 1974. Outline planning permission was granted in February 1974 without need for a public inquiry, only 1 year after the initial oil discoveries.

Although the proposed development was acceptable in principle detailed examination of the proposals and extensive consultations by the planning authority identified certain issues of concern, particularly visual, marine and socio-economic impacts.

UK planning legislation provides for certain matters to be reserved in the grant of outline consent and the Council formulated 18 such conditions. The conditions required further detailed submissions and approval by the Council, and were concerned with detailed plans for the terminal and its processes, precautions for safeguarding amenity and the scientific characteristics of the area, and the treatment of effluent and accidental oil releases.

The conditions provided feedback for Occidental's detailed assessment and design process, highlighting those issues requiring particular attention, and provided for continued control by the planning authority of development of the terminal. In addition the Council retained their own engineering and biological consultants for independent appraisal and monitoring of the development, and undertook studies of its socio-economic impact.

VISUAL IMPACT ANALYSIS AND LANDSCAPE PROPOSALS

Assessment and treatment of the visual impact of the terminal provides an excellent illustration of the means by which the procedures outlined above permitted application and integration of sophisticated and effective design and assessment techniques in areas of major concern.

The Orkney Islands have numerous areas of high scenic, scientific and archeological interest, of which several are of national or international importance. The low rolling landscape has long uninterrupted views across a predominantly agricultural setting of dispersed farms and houses. It is complemented by large expanses of water which reflect a dominant sky and cloud formations. The colors of pastureland, moorland, sea and natural stone in building and boundary walls seen in fequently changing light conditions introduce subtle variation and diversity into a landscape of great beauty, but one which lacks the visual enclosure so often created elsewhere by trees and hills. Construction of an oil handling terminal with large alien elements such as storage tanks in the predominantly flat and visually exposed rural landscape therefore presented a major visual design problem.

This was recognised in the conditions attached to the outline planning consent, 12 of which concerned environmental and design considerations, and embraced four main aspects of the development:

(i) detailed design and layout of buildings, plant and structure, including color and appearance, site boundaries and access roads, and the need for express approval for any installations or changes in type of process;

(ii) a landscape scheme for surface grading, earth mounds, screening trees, walls and fences, and their maintenance;

(iii) siting, design and layout of any temporary accommodation camp;

(iv) cessation of use of the land, removal of buildings, plant and structures, and reinstatement within 12 months of the end of operations on the site.

Stage III of the environmental assessment contained a detailed visual analysis, a design study and landscape development proposals.[5]

VISUAL STUDIES

The environmental consultants adopted an analytical approach to the problem of visual integration of the terminal. Analysis of the visual characteristics of the site and the site constraints used cones of site visibility to determine the major views into the site. The higher parts of the Island of Flotta itself provide a background from certain aspects and limits short distance views from the south. Dominant views were from the north.

The visual characteristics of the components of the terminal, which included storage tanks, process areas, buildings, single-point moorings and jetty, were determined. The visual criteria of profile, bulk, edge delineation, overlap of form, depth of field, and color were derived for description and analysis.[6]

The integrated assessment procedure related these criteria to the engineering requirements of the project which provide, with the site characteristics, the constraints within which the visual design exercise must operate. The scale, form and size of the tanks, and their relative horizontal levels were more or less fixed; conversely color was a relatively flexible factor.

Complete concealment of the terminal was impractical. Nevertheless, concern over potential visual impact required the maximum possible integration which would depend largely upon creative exploitation of those aspects of the design not rigidly set by technical or economic requirements.

Visual assessment and design was based on modelling of alternatives, using field survey and working models. Close liaison between the environmental consultants and the engineers allowed manipulation of the layout and position of the terminal components and their relationship with surface treatments, vegetation reinstatement and use of color to be developed and evaluated.

Visual impact of the terminal was minimised by locating its components so that they appeared to blend into each other and the landscape (Fig. 1). The 'blend and blur' strategy involved two aspects, treatment of structures and landscape development.

TERMINAL LAYOUT AND STRUCTURES

The first phase of terminal development for the Piper field required five 500 000 barrel crude oil storage tanks and two ballast water tanks of similar size. Subsequently two additional storage tanks of 950 000 barrels were added for development of the Claymore field.

The design solution reduced the visual impact of the shape and bulk of the tanks by compact grouping and overlapping the rows of tanks. The location of the groups and their overall elevation were adjusted to maximise the backcloth effect of the higher ground to the south, and surplus fill material was used to create landform on the exposed sides, which although not completely concealing the tanks further served to blend and blur their profiles. Oil processing facilities were similarly treated. While simple principles were involved the many combinations of tank design and layout, site location, view points, etc., made application of sophisticated computer techniques essential.

As a result of color studies and with the agreement of the planning authority, most components of the terminal were painted a subdued brown which blended with the landscape over a wide range of light conditions. To distract attention from the mass of the development certain structures were selected for contrasting color treatment (the 'isolation principle'[5]), e.g. operations and residential buildings, whilst storage sheds and workshops were the same color as the tanks.

LANDSCAPE DEVELOPMENT PLAN

The Stage III visual report[6] set out a preliminary landscape development plan which established basic principles for landscape treatment, conforming with the policies of the planning authority and the conditions attached to the planning consent. It also provided guidelines which were incorporated in detailed studies, designs and reports submitted to the planning authority.

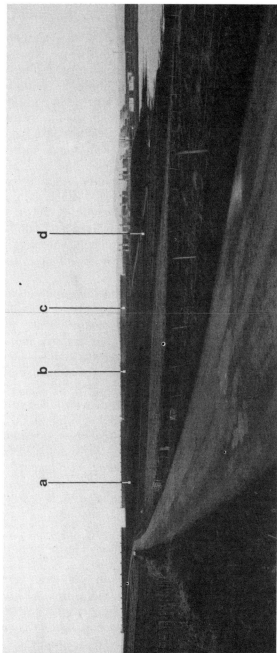

Fig. 1. View of Flotta (Orkney Islands) oil handling terminal from the south-east demonstrating visual integration. Note (a) use of land forms to obscure tank profiles; (b) grouping and overlapping of tanks; (c) subdued tonal treatment of structures; and (d) landscape reinstatement.

The development plan embodied five primary considerations. The use of ground moulding and landform is described above. Revegetation of the site was a major factor in securing maximum integration into the landscape, softening the terminal edges and differentiating the ground textures and colors to absorb the terminal structures. This included thicket and tree planting.[7] Attention was also directed to treatment of the space between tanks and buildings, and site component design; coastal reinstatement; and the production of a landscape management policy.

Consultations and discussions between officials and advisers of the planning authority and the developers, the Countryside Commission for Scotland, the Nature Conservancy Council and other agencies and interest groups, including the local Flotta Community Council, were an integral part of the decision-making and approval process by the Council throughout development of the landscape proposals. Extensive use was made of graphic representation, including 'before and after' studies to communicate ideas and proposals.

Although used here as an illustration, the landscape proposals were not confined to visual considerations. Other significant factors included requirements to combat erosion of disturbed ground and shoreline; basic reconstruction of surface conditions to support vegetation cover; safety, security and other operational requirements; and were linked to the environmental protection measures, such as containment bunds around the storage tanks.

The other major area of environmental concern identified at the outset, protection of the marine environment of Scapa Flow, was the subject of similar detailed studies, including those prepared by Dundee University on behalf of the planning authority and by the Heriot-Watt Institute of Offshore Engineering on behalf of Occidental, which contributed to the assessment, design and management procedures for the oil handling equipment (Stages IV and V)[8–13] (see also Johnston, these proceedings).

CONCLUSIONS

It is now some 8 years since the discovery of oil in the Piper field of the North Sea, and the subsequent initiation of the Flotta project. This review of the development process, highlighting the treatment of the visual design of the terminal, identifies several valuable lessons.

The project adopted a staged and integrated environmental impact assessment as part of a preventive environmental design strategy; the procedures were used to promote close liaison and cooperation between a responsive local authority and an environmentally committed developer. They provided a vehicle for the communication of intent and proposals, and were a basis for the resolution of matters of mutual concern.

The process has been notable for the speed of approval and development; this was aided by the studies by the local planning authority in anticipation of development and the ready response of the agencies advising the authority on many matters. The constructive use of planning conditions following outline approval highlighted matters of concern to the authority, and guided detailed design of the terminal.

The integrated assessment and design procedures provided a framework within which visual analysis and modelling techniques could make effective contributions. Both aspects were fundamental in tackling the visual design problem of integrating a large industrial development into the unique and fragile Orkney landscape.

The manipulation of the terminal components produced a compact layout, the impact of the elements was reduced by the loss of visual definition achieved through staggering and overlap, use of background, and artificial landforms. Color treatment aided integration of the structures with the landscape.

The adoption of a comprehensive landscape development plan has, by revegetation, surface differentiation, coastal reinstatement and management, further reduced the visual impact of both the terminal and associated construction facilities, and implementation of the final stages of revegetation with thickets and trees will further improve the environment. Attention to detail throughout the terminal by the coordinated design of buildings, roads and minor elements has contributed to the high environmental quality of the terminal.

The development at Flotta illustrates the value of an effective working relationship between the technical officers of the planning authority, the advisory agencies, and the technical staff of the terminal, with Occidental's environmental consultants acting as an authoritative point of reference to all parties. The committed and effective use of EIA procedures and the consultants' central role in providing advice, undertaking assessment and design, have served to raise the perception of the problems by all concerned, with an understanding of the alternative courses of action, in turn promoting the rapid implementation of a mutually acceptable solution. They have provided a framework for the effective and timely visual design of the oil handling terminal at Flotta, and its integration with the Orkney landscape.

ACKNOWLEDGEMENTS

Anil G. Veling, Simon R. Swaffield.

REFERENCES

1. Moira & Moira (1973). *Review of development plan interim strategy.* For Orkney County Council, Kirkwall, Orkney, UK.
2. Moira & Moira (1975). *Orkney structure plan* (revised 1979). For Orkney Islands Council, Kirkwall, Orkney, UK.
3. Occidental of Britain Inc. (1973). *A proposal for an oil handling terminal on the island of Flotta:* In consultation with Bechtel International Ltd.
4. Cairns, W. J. and Partners (1973). *Flotta, Orkney oil handling terminal. Report no. 1. Environmental assessment.* For Occidental of Britain Inc. and Associated Companies.
5. Cairns, W. J. Partners (1974). *Flotta, Orkney, oil handling terminal, Report 2, Visual impact appraisal and landscape proposals.* For Occidental of Britain Inc. and Associated Companies.
6. Aylward, G. & Turnbull, W. M. (1978). *Visual analysis: the development and use of visual descriptors.* DMG-DRS Journal.
7. Cairns, W. J. and Partners (1977). *Flotta plantations. Part I. Feasibility study.* For Occidental of Britain Inc.
8. Buller, A. T., Charlton, J. A. & McManus, J. (1974). *Potential movement of oil spillage and pollutants: Scapa Flow region, Final report.* Centre for Industrial Research and Consultancy, Dundee University, UK.
9. Halliwell, A. R., Johnston, C. S. & Ramshaw, R. S. (1974). *The effect on the marine environment of discharges into the sea from the Flotta project, Orkney.* For Bechtel International Ltd on behalf of Occidental of Britain Inc. by the Institute of Offshore Engineering, Heriot-Watt University, UK.
10. Buller, A. T. & Jones, A. M. (1974). *Criticism of the proposed site for Occidental's ballast treatment plant effluent pipeline.* For Orkney Islands Council, Kirkwall, Orkney, UK.
11. Jones, A. M. & Stewart, W. D. P. (1974). *An environmental assessment of Scapa Flow with special reference to oil developments.* Centre for Industrial Research and Consultancy, Dundee University, UK.
12. Bechtel International Ltd (1974). *Piper field development. Flotta terminal, Orkney. Marine facilities design and operations outline.* Vols. 1 and 2. In consultation with Occidental of Britain Inc.
13. Cairns, W. J. and Partners (1974). *Flotta, Orkney, oil handling terminal development, Report 3, Marine environment protection.* For Occidental of Britain Inc. and Associated Companies.

15
Whiddy Island, Bantry Bay

J. A. FEHILY
7 Clyde Road, Dublin, Ireland

ABSTRACT

Following closure of the Suez Canal in 1967, Whiddy Island, Bantry Bay, in south-west Ireland was selected by Gulf Oil for construction of an oil terminal from which super-tanker cargoes could be redistributed to European refineries. Gulf's contract with local authorities included provision for a landscape architect. Landscaping techniques employed in construction of the terminal included use of overburden to break storage tank silhouettes, paint color selection for storage tanks and a planting scheme to integrate the terminal with the rest of the island. Economic and social impacts have, through a combination of the Betelguese disaster and civil disorder in Northern Ireland, been most disruptive. To be successful in Ireland, developers must initiate and maintain dialogue with the local authority, the mechanism of which, with some differences, is similar to that in the United Kingdom.

INTRODUCTION

The history of Whiddy Island precedes North Sea oil and goes back to the Arab–Israeli war which resulted in the blocking of the Suez Canal. Shipping oil from the Persian Gulf in 60 000 ton tankers to refineries in Europe became too expensive. Gulf Oil looked around western Europe for a deep water port, where they could bring in 250 000 ton tankers, and redistribute the oil refineries on Antwerp, Milford Haven, Rotterdam and Spain. The choice was between the west of Scotland or the west of Ireland.

Bantry, in the county of Cork in south-west Ireland, was chosen because it had a massive deep water sheltered area, last used by the British Navy in 1937. It was fog free, closer to the refineries than Scotland and involved the minimum crossing of shipping lanes by the large tankers.

Whiddy Island is about 1–2 km from the mainland, west of the town of Bantry. When the Bantry project was being designed, the view from the surrounding coastal road system was a major consideration, since south-west Cork is an important tourist area. The County Council, Gulf Oil and the Irish Tourist Board had agreed that the visual impact of the project should be minimised.

LANDSCAPE ARCHITECTURE

The appointment of a landscape architect was written into the contract, although in contrast to the Flotta terminal (Sargent, these proceedings), four of the 12 tanks were built before the landscape architect was appointed. The terms of reference for the landscape architect were

  (i) to find a way of disposing of overburden;
 (ii) to prepare a color scheme for the tanks and pipe works;
(iii) to prepare a planting scheme which it was hoped would integrate the tank farm with the rest of the island.

Disposal of Overburden

The island was photographed from the surrounding mainland roads. The visual problem differed significantly from other projects discussed at this conference in that Whiddy Island is lower than the surrounding mainland. The tanks are never silhouetted against the sky. However, the tanks were, potentially at least, silhouetted against the water. Since the climate in south-west Cork is highly variable—cloud, no cloud, sun, cloud and shade mixed, blue water, grey water and endless combinations of these each day—it was apparent that the first essential was to build a bund between the two parallel rows of tanks using excavated material, making it high enough to obscure spaces between the tanks when viewed from the roads or the high land on either side.

Gorse (*Ulex* sp.) was indigenous to the island and might have been used to vegetate the bund but for fire hazard. Many other low lying shrubs and grasses were considered. Finally a grass-mix which would not die off in winter leaving a lot of combustible litter was found. It was slow growing and less than 9 inches high. Cutting the grass posed a maintenance problem. Machines could not be used on the bund. Cattle were considered, but the sides were too steep. Sheep grazing patterns were considered too unpredictable. Finally it was just left alone. The only other change necessary was to remove some of the old hedges to get a smooth transition between the tank farm and the rest of the island.

Color Selection

No precedent could be found which would help in selection of colors for a situation which varied from an almost moonscape scale inside the tank farm to views across 3 or 4 miles of water. Two basic techniques were used to answer the question of color.

First a series of perspective drawings were made recording the pattern of light falling on the tanks. From 10 a.m. to 5 p.m. the tanks were either in full light or full shade. But in early morning or early evening there was a sharp contrast between the areas of the tanks in shade which appeared black and the areas in sun which appeared white. This problem of reflection on a smooth surface, regardless of color, is almost insoluble.

Then all possible vegetation colors were photographed in both summer and autumn. On the basis of these colors 48 painted panels were made and set in the vicinity of the island. The panels were viewed against different conditions of ground, vegetation, sky and water background of the island. All but 12 colors were rejected. A series of cylinders of roughly the same shape as the tanks were made and painted with these colors. They were then tested around the island against the various backgrounds. In this way eight of the 12 colors were eliminated leaving a green, a green-brown, a blue and a blue-grey. These four colors were then chemically tested by the manufacturers and listed in order of performance, particularly the speed with which they would bleach. Because of extreme exposure in Bantry Bay this was an important factor. The tests indicated that the lighter the color, the more frequently re-painting would be necessary. Of the darker colors, the blue-grey fitted the sky, the sea most of the time, and the winter conditions, and was therefore the final choice.

Planting Scheme

One side of the island is extremely exposed to wind and the tank farm is close to this shoreline. On the other side there was more space, and it was also more important visually because the public road was closer to it. From the ordnance maps of the district, a series of plans were prepared showing all the visually vulnerable areas which had been identified on the ground. Relief profiles were measured along a series of east–west transects (prevailing winds are from the west) in order to identify areas of shelter behind which trees could grow. Superimposing the information from the relief profiles on the plans of vulnerable areas identified the potentially plantable areas (Fig. 1).

The planting was accordingly done on about 85 acres, using 90% indigenous materials (sycamore, ash, scots pine, thorn). In addition some contorta pine was planted, for rapid growth (Fig. 2). A year after planting, the sycamores and thorn were healthy but all the conifers (value about £20 000) were missing. During the winter they had all been eaten by rabbits and hares. It was necessary to re-plant, and protect all the trees from browsing. In retrospect, it is easy to see that on an isolated island with poor farming, the greatest number of livestock might be either hares or rabbits.

A good test of the landscaping work is whether or not there was adverse public reaction. In fact there was none, and that in an area where the local people are particularly vociferous.

SOCIAL AND ECONOMIC IMPACTS

Bantry is a small place, comprising a pier, the harbor, two streets which straggle up a valley, and before the oil terminal a population numbering about 1200. When the Gulf project started, over 600 highly paid workers were brought in, most of them Dutch, Swedish, German or British. The immediate result was that the town was shaken to its foundations and dying pubs boomed. Hotels and guest houses were built and the place just jumped like a mining town for 4 years. At the end of the project only 60 full-time jobs remained. That was in 1970, however, and it was believed that the infrastructure for a very good tourist industry had been created. South-west Cork had always been very popular with British tourists. As misfortune would have it, the Northern Ireland troubles broke at the same time and immediately British tourists vanished. Bantry carried on for the intervening 8 years with just the 60 jobs and a little tourism in the summer. Then, 18 months ago Betelguese blew up while unloading, wrecked the jetty and killed 52 people. The terminal was closed and Bantry was back where it had started 20 years previously.

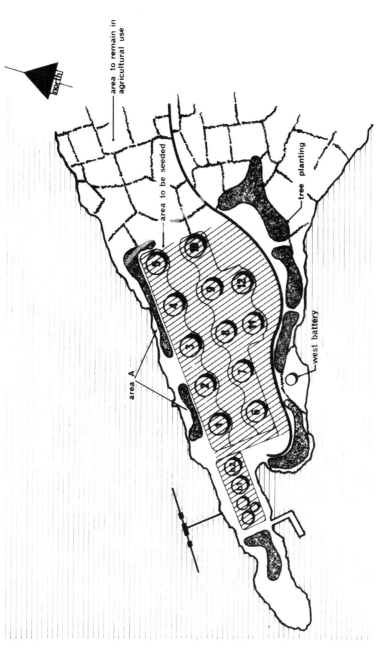

Fig. 1. Proposed landscape development of Whiddy Island tank farm.

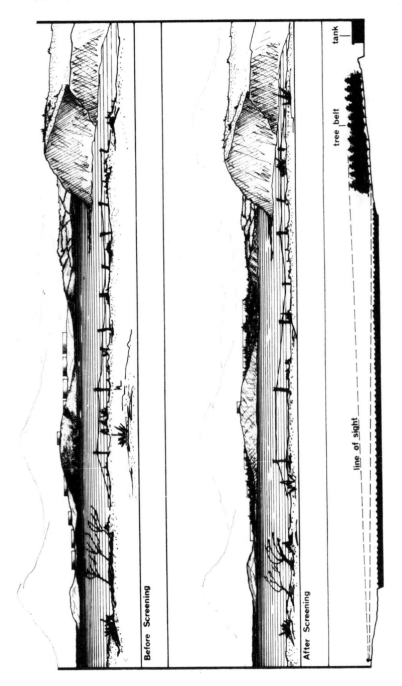

Before Screening

After Screening

tank

tree belt

line of sight

Fig. 2. Proposed tree screening of Whiddy Island tank farm.

LOCAL CONTROL OF DEVELOPMENT

The Irish local authority system is similar to that of the UK but there are certain differences. The chief executive of the county is called the County Manager and he is an extremely powerful man.

It is essential that prospective developers of oil-related projects meet the County Manager, the County Engineer and the Planning Officer when the project is first considered. One well informed company representative and a good local planning consultant should be present. The reason is simple. Ireland is a small place and people know each other, and they will not be intimidated by vast armies from abroad. There has been an unfortunate tendency, not deliberate, but arising from enthusiasm, to move in with every conceivable expert. The reaction is the opposite to that being sought. That first meeting should be quiet and the job described simply. How the job is to be administered, to the local authority and to the company, should be established. Administration in this case means who is going to talk to whom every day, exactly what information is needed and what form it should take. If possible the local authority should be persuaded to nominate one official to moniter the job all the way through.

Ireland also has a system, not present in the UK, known as the third party objection. Anybody in Ireland can object to any development, no matter where it is and who they are. Big projects and oil companies seem to be sitting ducks for the third party objection, so good, competent, local advice is absolutely necessary. The advice of the County Manager and the County Planning officer is particularly valuable, because at a public meeting they can tell which questions are genuine and which are not. In fact, in Ireland, the less you say and the more you let the locals say the less troubles you get into and the more successful you will be.

The job must be taken stage by stage with the local authority all the way through, and everything should be checked verbally before it is put in writing. This approach not only saves the project but also saves the local authority a lot of trouble as well. Developing companies will find it valuable to realise that if they have a legal obligation to a government department or local authority, it is just as serious as any other obligation and cannot be put off. This obligation must be slotted in, in the right way at the right time. The onus is on the company that has the resources to do it.

16

The Community Impact of North Sea Oil

H. A. GRAEME LAPSLEY
Orkney Islands Council, Kirkwall, UK

ABSTRACT

The effect of oil-related development on the Orkney community is discussed. Private legislation (Orkney County Council Act 1974; Orkney Islands Council Order Confirmation Act 1978) achieved wider control of oil-related development than that provided by existing statute. Development was controlled by planning conditions, contractual agreements and licensing, with minimal local authority participation in management and finance. An ex-naval anchorage was transformed into a major oil port with the physical development of the Flotta terminal. Regulations and precautions for environmental protection were required by Orkney Islands Council as Harbour Authority. Particular incidents, practical methods adopted to deal with them and the lessons learned are discussed. Social impacts include cirme, local employment prospects and effects on indigenous industry. Financial impacts include an anomalous rating situation and the effects of additional revenues accruing to the local authority.

INTRODUCTION

The impact of North Sea oil on any community depends on the type and location of the community concerned. In 1971 the population of Orkney was just over 17 000 with a gross product of £6·25 million. Today the population has risen by 1000 and the gross product has increased to over

£20 million. This figure does not include oil-related development and is based entirely on indigenous industries, principally agriculture, distilling, fishing, knitwear, tourism, boat building and a miscellany of smaller industries such as silvercraft, fudge making and electronic engineering. Thus without oil Orkney was prospering with a sound economy based on agriculture.

PLANNING LEGISLATION

In 1971 when oil exploration was starting, all the licensed blocks in the North Sea were closer to Shetland or to the Scottish mainland than to Orkney. It was therefore thought that Orkney would have advantage as a land base only for oil development in the Atlantic Ocean. Early in 1972, however, it became apparent that Orkney could be attractive for some oil-related development, and the Council gave consideration to Orkney's vulnerability under the existing planning legislation, which apart from two specific restrictions or conditions which could be attached to industrial development certificates was decidedly vague as to what other restrictions or conditions could be applied successfully.

The county of Orkney at that time did not require or want labor intensive development. There was very little unemployment and any major attraction of labor from the indigenous industries would have a very harmful effect on the long-term economy. The result was that Orkney County Council resolved to introduce private legislation to enable them to control development and to establish themselves as a statutory harbor authority. Shetland, because of the attraction of Sullom Voe as a site for an oil terminal, had already experienced land speculation on a major scale and had initiated a provisional order, later to become a substitute bill, some months before. Although Orkney and Shetland are a single parliamentary constituency, they do not form one local government unit. The two groups of islands are very different and Kirkwall and Lerwick, the county towns of Orkney and Shetland respectively, are separated by a hundred miles of open sea. The main provisions of the Orkney Act are

  (i) power of compulsory purchase over designated areas, i.e. those areas which the Council has designated as suitable for, and therefore probably attractive to, oil-related development;
 (ii) creation of two statutory harbor areas, with the County Council

as Harbour Authority with wide powers of control, particularly of marine works. This gives the Council power to transfer any surplus of revenue over expenditure from the harbors fund to a reserve fund, this fund to be used in the first instance to maintain the harbor areas, but also in any way considered by the Council as being of benefit to the community.

These powers have been of great value to Orkney.

AGREEMENTS WITH OIL INDUSTRY

While Orkney was engaged in private legislation, the Occidental Group (North Sea Consortium) advised the Council that they were looking for a landfall for a pipeline from the Piper field and that among other places they were interested in the possibilities of Scapa Flow. Shortly after, the group came back saying they would indeed like to bring their pipeline into Scapa Flow and to establish an oil terminal. Their first choice did not appeal to the Council and they were quickly persuaded to develop on the island of Flotta.

Flotta had an ageing population of 80 people, the best land was Grade C and little of it, and there was little hope of Flotta ever becoming a viable economic unit under normal circumstances. Although Orkney did not then have its statutory powers Occidental accepted them in principle and readily agreed to conduct negotiations as though the Orkney County Council Act was in force. Accordingly, with the Council's blessing, and that of the District Valuer, Occidental negotiated with the land owners concerned in Flotta, established a price per acre and, having purchased the land, sold it to the Orkney County Council, thus avoiding the lengthy procedure involved in compulsory purchase. So, as in Shetland, the Council owns the ground but there the similarity ends, in that Shetland exercises control not only by additional agreements but by participation on a large scale. Shetland has licensed and constructed the terminal facilities at Sullom Voe, requiring very large sums of money, whereas Orkney has proceeded on a course of control by license and agreement involving a minimal capital investment, bearing in mind the small population available to carry such a debt. However, circumstances in the two island groups were quite different. Orkney was faced with four oil companies acting as a group, thus simplifying negotiations, whereas Shetland had to deal with many more companies. The agreements which we have with the Occidental Group are

    (i) initial planning consent, which contained 17 important conditions;

    (ii) lease-back of ground on which the terminal is constructed to Occidental, in which the relevant planning conditions are repeated, in addition to those conditions laid down by the Council as landlord;

    (iii) licenses for marine works in respect of: single-point mooring towers; underwater oil loading and ballast discharge pipes in connection therewith; LPG jetty; incoming oil pipeline; treated ballast water discharge line.

Each of these licenses contains strict conditions and may be terminated by the Council should there be failure to observe those conditions. The Council may also insist on joint usership, should this be deemed practical and desirable.

In addition there are other important agreements. The first is a Disturbance Agreement, based on throughput of fiscalised crude oil. The second is the Harbours Payment Agreement, Part 1 of which provides that all harbor costs, headquarters' building, radar, marine officers, boat crews, boats, salaries and wages, etc., will pay to the Council a fixed sum for the first 4 years, after which the additional sum will be directly related to the oil shipped. This payment was governed by a complex formula intended to compensate for inflation. However, it proved so complicated that it was unworkable and a new simpler formula based on the retail prices index has been negotiated. During the period of fixed payments, provision is made for an additional levy on any oil exported above 20×10^6 tons in any one year, and a separate payment is levied on LPG.

Having exercised its statutory powers to enter into joint ventures, the Orkney Islands Council is the controlling partner of the Orkney Towage Company Limited. This company, which operates three modern ocean-going fire-fighting tugs, is the only towage company licensed to operate in Orkney waters and so contributes significantly to local revenue. The general harbor byelaws provide that a tug or tugs will be in attendance on all vessels designed for the carriage of petroleum, LPG or dangerous goods, and the 1978 Orkney Consolidation Act provides that only those tugs licensed by the Orkney Islands Council shall operate in the statutory harbors.

As well as being Harbour Authority, the Council is also Pilotage Authority and pilotage is compulsory in Scapa Flow. To oversee the statutory harbor we have an up-to-date Department of Harbours, manned by marine officers, all of whom are well qualified and experienced, with a minimum qualification of master with foreign going certificate. These officers have three main duties: normal headquarters watch, surveillance and pilotage. When a tanker is loading, she is inspected at least twice in every watch and the Director of Harbours has full authority to control loading.

Strict ecological control stems from a detailed pre-development study by the Orkney Marine Biology Unit of Dundee University, which is permanently employed by the Council on monitoring the water and shores of Scapa Flow. Occidental have been most cooperative in this and have done all they can to maintain the high standards demanded of them; in addition they employ environmental consultants. The procedure for loading crude oil and deballasting are as foolproof as words can make them, but the human element calls for constant vigilance to prevent spills. There is permanent worry about possible oil spills and it must be accepted that in any terminal some minor spillage will occur because of human error. Nevertheless, if proper precautions are taken and proper discipline enforced, it is unlikely that such spillages will be of much consequence.

It must be accepted, however, that there is no way of providing complete cover for a major disaster, although all possible steps are taken to prevent such an occurrence. No tanker can enter, leave or manoeuvre within Scapa Flow unless it has a fire-fighting tug with adequate bollard pull in attendance. Patrol launches operate by day and night and a searchlight on a loading tanker can have salutary effect.

The following figures demonstrate the effectiveness of these methods. Since shipments began at the Occidental terminal on 11 January 1977, 804 ships have been loaded, of which 683 were tankers carrying 400×10^6 barrels of crude oil. The other 121 vessels were liquid gas carriers. During this period ten minor incidents occurred, resulting in an aggregate spillage of 60 barrels. It is agreed that this is 60 barrels too many, but a spillage ratio of 0·000 019 5% is proof of the effectiveness of the preventive measures taken.

The necessity to impose the strictest control cannot be over-emphasised and success must never lead to complacency. Prevention of oil spillage is a continuing worry and expensive. But it is a great help when, as is the case in Orkney, the terminal operators share the worry and the expense.

SOCIAL AND ECONOMIC DISADVANTAGES

The first and most obvious impact of North Sea oil is perhaps the environmental problem which any major installation presents, and this problem
is emphasised in the case of oil, because it is an emotive subject. To some
people it means money. To others it means potential mess, and pressure
groups for and against are very much in evidence.

Having taken into account the environmental problems, economic
conditions will normally determine the situation of a terminal and these
economic conditions are of vital importance to us all. On the other hand,
there is no reason for a properly sited terminal to be unduly obtrusive. At
present on Flotta we have five 500 000 barrel tanks and two 1 000 000
barrel tanks, none of which is higher than 46 ft, which means that they
blend into the landscape because the hills of Hoy act as an effective background. In fact, Flotta is as near a thing of beauty as an oil terminal can
be and it has won international acclaim as a model of what can be
achieved by a good relationship between a local authority and a developer
(see Sargent, these proceedings).

Employment

On the debit side there is no doubt that the presence of a highly paid
nucleus of workers in a small community has a disturbing effect. For
example, during the construction phase extremely high wages were paid,
which meant in turn that local building contractors had to increase
bonuses to retain their men, which in turn increased building costs. In so
far as permanent staff is concerned, there has been very little noticeable
effect in Orkney and while most industries have lost one or two men, there
has been no drastic enticement of labor away from indigenous industries.

Crime

Despite the influx of construction labor, which reached a peak of 1700,
their impact was scarcely felt because, before starting, the developers
were advised to make arrangements for their labor force to be sent south
for their rest periods. For this reason, and because the construction was
on an island, few, if any, of the construction workers came to the mainland of Orkney, except *en route* to the airport. In Shetland, however, the

Sullom Voe development is on the mainland and therefore the workers can easily move into Lerwick, Scalloway and other places.

To illustrate the result of these different situations, the 1976 figures for crimes and offences made known were exactly the same (1374) in Orkney and Shetland, but in 1977 in Shetland the number had risen to 3103, whereas in Orkney the figure had fallen to 1195. In 1978 there was an increase of 13, followed in 1979 by a decrease of 101, which indicates that oil-related development has not affected crime in Orkney.

So, in general, where geographical and physical conditions confine a workforce to a limited area, it is unlikely that the crime rate will be affected, but where a highly paid labor force living away from home has easy access to the community extra policing may be required.

Rates

When an area attracts a major development, the populace is entitled to expect that the significant increase in rateable income will be to the benefit of the community. In Orkney, however, a problem arose because of two factors. First, at the Valuation Appeal Committee the terminal operators were successful in obtaining industrial de-rating for the whole terminal, although in fact they had only expected that this would be applied to the processing area.

The second point stems from revaluation. One of the criteria used by the assessor in valuation is the rental which a property might reasonably be expected to obtain. In Orkney at the last but one revaluation, rentals were very low but between then and the current revaluation, oil-related personnel had arrived and rentals had increased tenfold. The net result was that new valuations in Orkney represented 4·9 times the old, which is much higher than the national average. It also meant that Orkney's standard penny rate product (which is arrived at by a central government formula which determines the amount of the resources element in the Rate Support Grant) rose to a level which resulted in the withdrawal of this grant element, but ignored the fact that because of industrial derating, the rates received from Flotta did not make up for the complete loss of our resources element (£1·5 million).

This matter has been satisfactorily resolved by mutual agreement between the Council and the terminal operators. Nevertheless, it is a very important factor to be taken into account when faced with major development. In a small community the cost of providing the infrastructure might be more than the resultant increase in income from rates.

A final observation on the side of disadvantages is that one of the most difficult features of such a development is to keep the public informed, as oil companies are notoriously reticent about their intentions.

SOCIAL AND ECONOMIC ADVANTAGES

Additional Revenue

Orkney derives additional revenue from its Disturbance Agreement, its Harbours Agreement, the sale and lease-back of Flotta, and the profits from the Orkney Towage Company Limited. This revenue, with the exception of the lease-back rental, is channelled through Harbours and goes into the Reserve Fund. It is then allocated under three headings:

    (i) Development Fund   80%
    (ii) Community Facilities and Conservation Fund   10%
    (iii) Leisure and Recreation Fund   10%

Under these proposals, the annual surpluses on the harbor account have been allocated to each of the funds in accordance with the stated percentages. Future allocations will be the subject of annual review.

Development fund. The main object of this fund is to ensure Orkney's long-term economic viability. To this end the Islands Council has used or committed £564 000 to development, including the purchase of the controlling shareholding in the Orkney Towage Company Limited, and assistance to local fishermen to obtain new boats by purchasing a share in these boats. Many other projects are being undertaken, including the provision of a new abattoir and meat processing unit at Kirkwall, a slaughterhouse in the northernmost island, and development of an existing fish processing factory in one of the north isles. Farmers and small businesses have been helped by interest subsidies on bank loans and in addition, the Council has purchased the ex-Ministry of Defence base at Lyness, which, with its deep water quay, may well attract development to the island of Hoy.

Community facilities and conservation fund. Each Community Council in Orkney has so far received three grants of £1000 from this fund, which with an annual subsidy based on district population totals £80 000. £10 000 has been donated to a gallery of modern art; £38 000 has been contributed towards the installation of domestic mains electricity supplies in the islands of Hoy, Westray, Papa Westray and Egilsay; £135 000 has been distributed to old age pensioners as Christmas bonuses; a toy laboratory for handicapped children has been established, and a conservation fund for buildings and sites of architectural and historical interest has been set up.

Leisure and recreation fund. This fund is used to provide permanent facilities for leisure and recreation. The aim is to encourage provision of sporting and other recreation activities in such a way that members of the community will have a wide choice of leisure pursuits in which young and old people will be encouraged to participate. Benefits will also accrue to the growing tourist industry, which now contributes 10% of the Orkney gross product. £26 000 has been given to provide community facilities and local sports organisations have so far received £34 000.

The figures are not much in national terms, but Orkney is a small community and it is already benefitting directly from the establishment of the reserve fund.

Employment

The credit side of this subject is, of course, much larger than the debit side. On Flotta 288 permanent jobs have been created, of which 230 are filled by Orcadians, many of whom have returned to Orkney. Of 72 personnel involved in the catering and domestic services on Flotta, 47 are Orcadians. On the marine side the transport and rope-running launches employ 88, of whom 80 are Orcadians, and many of whom have returned after service at sea. The Harbour Authority of Orkney Islands Council has a staff of 32, including 15 marine officers, and 12 launch crew and at the moment 34 Orcadians are employed by the Towage Company. Additional job prospects provide an outlet for at least a few school leavers, who would otherwise have had to seek skilled employment outside Orkney.

Local Benefits

Flotta was an island with a small ageing human population, poor land and limited prospects. Today it has new housing, a school which has had to be extended, improved community facilities, an improved water supply, an air strip, improved piers, better roads, mains electricity, a better ferry service, and local employment prospects. The benefits far outweigh the disadvantages and Orkney looks forward to a long, cordial and prosperous relationship with the oil industry.

CONCLUSIONS

It is difficult to apply Orkney's experience precisely to other areas, in that Orkney is an island authority, which means it is a single tier, all-purpose authority, with responsibility for district and regional functions. Valuation is joint with Shetland. Orkney is fortunate in having its private legislation, which gives it good bargaining power with potential developers, but nevertheless I believe that many of the lessons learned can be adapted to suit particular requirements of other areas.

It seems essential, particularly for smaller communities, to have vetting machinery for preliminary consideration of applications for major development in their area and to determine whether such an application conforms to their policy. The policy will vary according to the needs of the community and will depend on the type of development. For example, is it labor intensive; if so, is that what is wanted; is there an infrastructure appropriate to such a development; if not, can the community afford to provide one; would it be better for the general economy to channel such an application to another authority if need be, if it is felt that there is a suitable infrastructure in an area of high unemployment? These are all political questions which each community must adapt to local conditions.

It is equally important to make the position clear if there is any particular industry which is not wanted. For example, Orkney County Council stated clearly in their interim strategy that concrete platform building was not acceptable in Orkney. In the structure plan the Islands Council stated that they did not want a refinery in the islands area and more recently, they have made it clear that they do not want mining for uranium ore in the islands.

However, having decided that the proposed development is desirable, great care must be exercised in appointing negotiators. One cannot negotiate successfully by committee. In Orkney and Shetland the initial negotiations were done by the former general managers, who subsequently became chief executives. Although the Shetland Chief Executive left before the job was complete, chief executives are generally more permanent than elected members. Even in the small context of Orkney, 42 councillors have come and gone in the last few years.

To empower a suitable official to carry out initial negotiations is not an abrogation of power by the elected members, provided arrangements are made for the official to report periodically to a committee, probably policy and resources or the equivalent. That has been the procedure in Orkney and the Chief Executive has always gone to committee with the penultimate draft of any agreement, having of course made it clear to the developers that everything was subject to approval by the Council. The system has worked, because by the penultimate draft stage a great deal has been learned about the other side's thinking and it is possible to guess fairly accurately how far they will be prepared to concede points. The message is not that negotiating powers should be inevitably or invariably delegated to an official. In fact in a politically dominated authority this would be inadvisable, but in the end, negotiations must be carried out at a personal level, which can only be achieved by keeping the negotiating party small.

It would probably be difficult for other non-island authorities to obtain such wide statutory powers in today's political climate but, nevertheless, if there is any doubt of the adequacy of existing powers, then these should be strengthened by any means to increase the bargaining power of the local authority. If there is a prospect of setting up a new and commercially viable harbor, then the local authority should take powers to control it as Harbour Authority. Many conservancy boards and dock boards today have enormous reserves but have little or no power to spend these reserves on anything but harbor installations. They cannot apply balances to the benefit of the community as a whole.

Agreements should be long-term and written in clear unambiguous language. Oil companies have a surprisingly large turnover of staff and it is a time-consuming job dealing with newcomers who were not involved in the initial negotiations and consequently try to score points by arguing about words rather than maintaining the spirit of the agreements. Encourage incoming staff to set up house in established communities and not to concentrate in particular localities. In Orkney incoming permanent staff have integrated amazingly well into the community and are to be found in nearly every part of the Orkney mainland, and most of them make an active contribution to community affairs.

Experience with oil companies in Orkney and elsewhere would seem to indicate that they are anxious to acquire the good will of the community and, if given the chance, to play a part in community affairs.

17

The Impact of North Sea Oil on Stavanger

K. E. EGELAND
Rogalandsforskning, Stavanger, Norway

ABSTRACT

During the 1970s oil-related activities in the Stavanger area increased rapidly, employment rising from 500 to over 14 000. This development was in an area having about 140 000 inhabitants and fairly modern export oriented industry. Until now interest has largely been concentrated on ways of coping with the rapid increase in employment and on methods of channelling some of the expansion to other areas. The long-term implications of oil-related activities on the local community and industry should now receive greater emphasis. An oil policy relevant to their needs must be focused on integration, increased competence, stability and independence. The internal structure of the oil industry must therefore be better understood, particularly the organisation and strategy of oil companies.

INTRODUCTION

Oil-related activities have affected the Stavanger area in many ways. This is obvious from the increase in oil-related employment which climbed from 500 to 14 000 during the 1970s. This has occurred in an area having about 140 000 inhabitants. Of those, 14 000 are now oil-related employees, 10 000 of whom live in the county of Rogaland and approximately 1400 of whom are foreigners. It is not known how many oil-related employees live in the Stavanger area, but the figures are probably fairly close to those for the county as a whole.

Oil-related activities have produced a series of economic, social and cultural effects. But because the activities have been in an area fairly densely populated for Norway, it is hard to identify simple cause-effect relationships and effects influencing the area as a whole. The different parts of the Stavanger area seen from a geographical, economic, social or cultural point of view have been strongly affected by oil activities, but in different ways. Some of the effects are easily seen, others are not. Some effects occur quickly while many, and perhaps the most important, can be measured only after several decades.

One obvious effect is that Stavanger has become the national center for oil activities. The public debate, the authorities and the news media have largely concentrated on the size and growth of oil activities, and whether or not the Stavanger area can cope with the expansion. This is an important question, but too strong a focus on it has probably diverted attention from the character of different parts of the oil activities and their long-term implications for the area. The Stavanger area has increasingly adapted to oil activities and the area mirrors both the positive and the negative aspects of oil-related growth in many ways.

It is important to understand in more detail how different oil activities affect the local community in order to develop central and local government machinery and policy for planned control of development. Now is the time to question what will happen in the Stavanger area when the production from the fields on the Norwegian continental shelf starts to decline. It is hoped that the area will have a life after the oil activities have gone.

Two subject areas are of particular importance when considering the impact of oil on the local community. First is stability and second is integration, particularly in provision of goods and services from other parts of the economy, the recruitment of personnel and development of expertise and special knowledge. The oil industry should also be evaluated on potential short- and long-term effects on the local economy and mutually dependent relationships which have arisen or may arise between the oil industry and the Stavanger area. Such an analysis is preferably done from two different angles simultaneously, first on the characteristics of the various parts of the oil-related activities and second on the attitudes and behavior of different oil-related companies and institutions.

STABILITY

All branches of oil-related activities, except drilling, catering and construction of platforms in 1976–77, have so far shown steady increase. Long-term stability, however, depends on the internal stability and continuity of the different parts of the oil activities under development.

Oil-related activities are normally divided into exploration, construction and production phases. Administration such as main office activities and activities of governmental agencies are involved in all three phases.

Administration and production are the two categories of activity which have the greatest stability. Production from an oilfield extends at least 30 years, while an exploration or construction program lasts relatively few years. If more long-term stability is to be established in the latter activities then a policy must be established which will ensure that when one exploration or construction program is finished, a new one will start.

The Stavanger area is the center for most of the oil companies and for the Norwegian Petroleum Directorate (NDP). It is also the operational base for the Ekofisk, Frigg and Statfjord fields. Thus oil-related activities in the Stavanger area are characterised by stability. Oil-related development in north Norway will be concerned mainly with exploration and perhaps later, if commercial fields are located, on construction and the directly operational parts of production. Activities in north Norway are thus more characterised by instability.

A more detailed analysis of stabilisation can be made by examining the companies and personnel involved. The NPD, Statoil, and foreign oil companies which have established their Norwegian headquarters in the Stavanger area are the most permanent part of oil-related activities. They have invested large sums in exploration and development of several fields, so are most unlikely to move. However, oil companies hire a variety of specialist companies for limited periods to do specific tasks or jobs. Without new contracts these companies' presence in Stavanger may terminate or decline. Even if new work follows a completed contract, the oil company may choose to hire another contractor. This relationship between oil companies and the contractors is repeated at the next level, i.e. between the contractors and the subcontractors, and so on. The organisational structure thus means that oil-related activities support less stable employment than total figures might indicate.

Personnel working in the oil industry are to a great extent influenced by personnel policies, common in many multinational companies, which result in employees moving between different offices and branches of the companies frequently.

Shortage of expert personnel leads to many individuals moving from company to company, going one step up the wage ladder at each change. One effect of these changes is that foreign personnel have found problems from Norwegian labor relations practice. Local authorities have suffered the disadvantage that management in the foreign oil companies changes frequently, making it difficult to establish mutual understanding on development.

Stability is further affected by the location of oil company head offices. Development of oil-related activities in the Stavanger area implies strong increase in the proportion of outside control of the economic life of the area. The sole task of affiliates operating in Stavanger is to optimise business on the Norwegian continental shelf. Decline in oil-related activities will be immediately reflected in those companies' employment. If either head offices of the parent companies or the owner interests were located in the Stavanger area, the chance of a reorientation to other markets or products when the oil activities on the Norwegian continental shelf start to decline would probably be greater.

INTEGRATION

Branches of the oil industry which according to official statistics (August 1978) have over 1000 employees are drilling, production, transportation to oilfields, technical services and construction of platforms onshore and offshore. The probable future pattern is that employment concerning oil production will rise while employment connected to drilling will probably decrease.

Although administration and production have positive stability, they also, perhaps, make least direct demand on other sectors of local business life. All the base and operational activities connected to the exploration and development of an oilfield, however, increase demand for goods and services from the local economy. In this way development of the oil activities may increasingly isolate the oil industry from other parts of the local economy. Perhaps the greatest impact will be on the public sector and general service activities, sectors which can hardly create new activities themselves.

Of those totally involved in the oil industry in 1974, 87 companies employed about 4500 people; by 1978 there were 187 companies employing about 13 500 individuals. On the other hand, of those companies only partially involved in the oil industry, 20 employed about 300 persons in 1974 while the number of such companies had increased to only 39 in 1978, with an employment of about 700. These statistics are only approximate, but they provide a clear indication of the oil industry's relatively low integration with companies established in Stavanger prior to oil.

The head offices of most parent companies are outside the Stavanger area, but the potential for developing a stable contact net with the rest of the local economy would be better if they were in Stavanger. At present many companies which have established themselves in Stavanger already have an international contact net with the other companies with which they do business. Foreign contractors also find Stavanger unsatisfactory for obtaining special goods and services. The impression is that in many instances they would have bought on the local market, but they are unfamiliar with it and will not take the necessary time to rectify this. Instead they buy most of what they need through the established international network.

The stability of the activities specifically concerned with administration and production contributes to the integration of employees with the local community, at least for Norwegian employees. Foreign personnel are normally more transient and in this sense will not take part in the integration of the oil industry. However, Norwegian personnel are increasingly being recruited, which will strengthen the process of integration.

The degree of integration is also reflected in the attitudes of foreign companies and their personnel. Interviews with representatives from foreign service companies reveal a fairly hostile attitude to Norwegian authorities and policies. It is often felt that Norwegian authorities hinder company activities and that taxes and costs are too high. These factors create, as the companies see it, problems in getting qualified personnel.

Nevertheless Stavanger seems to be rated a good place to live, particularly for children. Stavanger has now an English, an American and a French school, which are said to be of a high standard.

EXPERTISE AND TECHNOLOGY

The oil industry is a new element in the economic life of Stavanger and Norway, implying that the area has attracted companies, experts and skills which did not exist there before. This implication is reinforced by the oil companies and the NPD having their main Norwegian offices in Stavanger. The expertise within petroleum technology is perhaps the most important one for Stavanger which has now a solid population of geologists, geophysicists, reservoir engineers and so forth. One result of this is that before the oil industry started, the possibility of getting a relevant job in Stavanger would be limited for young people from Stavanger with university degrees; but opportunities are now much better for those with a background in the natural sciences, law and economics.

Much of the technology required as a result of oil industry is fairly simple. The important and specific expertise is most probably concentrated in petroleum technology and the management of exploration, construction and production. In addition, of course, basic knowledge in a wide range of subject areas is necessary to tackle the wide variety of jobs generated by the oil industry.

Foreign companies will normally have their top managers and experts at the head office, e.g. in Houston. For instance, the most thorough analyses of the exploration data and reservoir simultation studies are normally done at the head office. The Stavanger office will largely be restricted to an operational base, functioning under the direct control of the head office. It is from the head office that experts are sent to Stavanger when specialised jobs or specific problems arise. This is reflected in the fact that until now, almost no research and development by the foreign companies has taken place in Stavanger. The training of personnel is also limited. This is particularly true for foreign contractor companies.

The pattern of development of expertise is also reflected in the nationality of personnel in the various positions in foreign companies. Seldom are managers or senior personnel in the various technical departments Norwegian. Above a certain level, the companies recruit personnel from their worldwide organisation, making it difficult for Norwegians to establish a career in these companies. Thus, to a great extent, top Norwegian expertise must be obtained through Norwegian companies which can recruit Norwegian personnel working in foreign companies, or they can employ foreigners and foreign companies as experts and learn the business and the technology the hard way, by their own experience.

DEPENDENCE AND FUTURE POSSIBILITIES

The relative stability of the oil industry means that the local community's dependence on it is not yet a pressing problem. Nevertheless the economy of Stavanger is increasingly adapting to this new, dynamic and resourceful economic activity, which the oil industry both is and generates.

A real problem can be foreseen, however, when the oil-related activities start to decline, and only then will the extent of Stavanger's dependence on the oil industry be fully realised. It is thus necessary to examine such problems now and try to find the means to establish a 'way-ahead' policy. By so doing we also focus on the characteristics and the qualities of the oil industry in the period when it will be an important factor of the economy.

The important question therefore is just how dependent are the oil-related companies in the Stavanger area on the activities on the Norwegian continental shelf. If the companies are mainly foreign and directed towards the operation of a specific oilfield, the dependence is likely to be very high. On the other hand, specialised technical and service companies with head offices also in the Stavanger area probably have considerably more flexibility.

A further element in examining the problem of dependence is the extent to which the existing organisational structure and personnel can be applied to new activities. What kind of impact or potential for the future is there in the fact that Stavanger is one of the cities in Europe with the highest number of petroleum geologists? The same question might be raised for other categories of personnel and for the various branches of the oil industry. For example, up to the 1960s Stavanger had a canning industry. Then the fish vanished and so did the industry. The canning industry, however, had needed labels for their tin-boxes. This was an important element in the development of a thriving printing and graphics industry in the Stavanger area today.

In looking at the oil-related activities in the area and how they will develop one might expect that the focus would be on administration and operational activities connected to production from different oilfields. This may not, perhaps, be the best basis from which to orient to new markets and products after the Norwegian oil story has come to an end, although it might be easier for specialised companies involved in exploration manufacturing and construction activities.

CONCLUSIONS

The focus of attention should now be shifted from the quantitative to the qualitative aspects of the oil-related activities. The qualities and potentials of the oil industry have not yet been developed and exploited in the most efficient and rational way. The stability, integration and expertise is less than figures alone might indicate, mainly due to factors concerning the organisational structure. The characteristics of the various parts of the activities and organisational structure must be examined in more detail. In the long run, it is unsatisfactory to let the activities and the company structure mirror only the phases of the development of oil activities.

If Stavanger wants to have an oil industry after the petroleum on the Norwegian continental shelf has been exploited, that industry will have to become international. Important and strategic questions then have to be raised concerning which activities Norwegian companies should focus upon. It would be unwise to establish Norwegian expertise in all oil-related areas.

A planned division of labor vis-à-vis foreign companies should be established. An important aspect of this is the establishment of Norwegian expertise which later on can be sold on the international market. A planned effort to develop research and education in the Stavanger area would be a natural consequence. It is irrational if the government tries to hinder development of these activities with the argument that the expansion of oil-related employment must be controlled. In reality the reverse is more the case. Because the development of oil-related activities is to a great extent dominated by foreign companies and experts, there is a pressing need to develop Norwegian expertise as fast as possible. Establishment of such expertise through research and education in Stavanger will have the additional and important benefit of being able to draw on the expertise of the existing oil industry.

BIBLIOGRAPHY

Anon. (1973–74). *Petroleumsvirksomhetens plass i det norske samfunn.* Storting-smeld No. 25. Oslo, Norway.

Anon. (1977). *North Sea oil: the application of development theories.* Institute of Development Studies, University of Sussex, UK.

Anon. (1978–80). *Petroleumsundersolkerser nord for for 62°N.* Stortingsmeld No. 46. Oslo, Norway.

Anon. (1979–80). *Om virksomheten pa den norske kontinentalsokkel.* Storting-smeld No. 52. Oslo, Norway.

Egeland, K. E. (1978). *Norge og oljen—Stavanger ble stedet.* Rapport No. 3. Rogalandsforskning, Stavanger, Norway.

Godo, H. (1978). *Noen trekk ved befolkningsutvikling og flyttebevegelse pa Nord-Jaeren 1967–76.* Notat No. 1. Rogalandsforskning, Stavanger, Norway.

House, J. D. Adapting to Aberdeen: the oil companies' perspective. In conference on *Oil and Scottish Society.*

Hunt, D. (1977). The sociology of development: its relevance to Aberdeen. *Scottish Journal of Sociology,* **1** (2), 137–54.

Hunt, D. & Atkin, A. (1979). *Oil-related impact on the industrial infrastructure of*

Aberdeen. Report to the SSRC North Sea Oil Panel, University of Glasgow, Glasgow, UK.

Moore (1979). *International register of research on the social impact of offshore oil development.* 3rd edn. Department of Sociology, University of Aberdeen, UK.

NAVF-RFSP (1978). Samfunnsmessige konsekvenser av petroleumsvirksomheten. Hva kan vi laere av erfaringene fra Sor-Norge? In *Forskningsrapport,* ed. K. Stenstadvold. Bergen, Norway.

Nodland, S. I. (1980). Stavangeromradet og oljeverksemda—vekst og utvikling eller bare vekst? In *Regionale erhvers—go arkeidmarkedsanalyser i Norden under 70' arenes okonomiske krise.* Rogalandsforskning, Stavanger, Norway.

Stangeland, P. (1977). *Condeep-En arbeidsplass i Stavanger.* Report No. 1. Rogalandsforskning, Stavanger, Norway.

Stangeland, P. (1980). *Hva skjer med Stavanger? Sosiale konsekvenser av oljeindustrien.* Universitetsforlaget.

Stenstadvold, K. *Arbeidsmarkedskonsekvenser av oljeaktivietetne i Sotra/Bergen—regionen.* Report No. 8. Industriokonomisk Institutt, Bergen, Norway.

Stenstadvold, K. *Arbeid og oljeaktiviteter i Sotra/Bergen 1977–78.* Arbeidsrapport No. 11. Industriokonomisk Institutt, Bergen, Norway.

18

The Impact of North Sea Oil on the Grampian Region and Aberdeen

T. F. SPROTT

*Department of Physical Planning, Grampian Regional Council,
Aberdeen, UK*

ABSTRACT

The economic and environmental character of the Grampian Region, its settlement pattern and the role of Aberdeen City as the offshore oil and gas 'capital' of Europe are outlined. Pre-oil (post-war to 1970) economic decline is described. During the 1970s exploitation of North Sea oil and gas had significant social and economic effects; extensive development of infrastructure was required. Gas pipeline landfalls, separation facilities and the prospect of petrochemical processing appeared. Authorities were presented with problems and opportunities for economic forecasting and forward planning. In the 1980s North Sea exploitation may be at different scales and involve a new economic and social change. Oil and non-oil industries have diverging prospects; oil-related employment will increase, affecting the rural economy, and the need to expand infrastructure in certain areas will continue. There will be opportunity for science-based industries, for greater community support by the oil-majors and continued need for financial support from central government. Thus the 1980s will see new planning opportunities and continued need for coordinated central and local government policies. The oil and gas industries promise to be both economic catalyst and social cushion. The community faces a variety of risks should the economy become seriously unbalanced.

INTRODUCTION

The Grampian Region is the third largest mainland region in Scotland, covering some 3400 square miles, with a population of 470 000, almost 50% of whom live and work within 20 miles of Aberdeen City. A further 25% live in 34 small settlements with populations from 1000 to 19 000, and the remainder are in thinly populated upland areas. The region is varied in both its physiography and its economy: uplands, including the Cairngorm mountains and winter sports areas, in the south-west; a lowland coastal plain, rich in agriculture and forestry; a long and environmentally sensitive coastline with many harbors, and markets and processing bases for the fishing industry. Aberdeen, one of Scotland's four major cities, is the administrative center and economic focus of the region and has long been important as a center of education, scientific research, manufacturing and service employment.[1-3] It is now renowned as the European offshore capital for the North Sea oil and gas industries, servicing oil company exploration and production activities from bases in and around the city.

The effects of major economic change resulting from the development of hitherto unforeseen natural resources are not new to the Grampian Region and Aberdeen. During the 18th and 19th centuries comparable periods of economic growth resulted from development of the textile, fishing, shipbuilding and, later, the granite industries. However, the region has enjoyed no change comparable in scale or effect during this century to that carried by oil, and is fortunate in having attracted and developed those aspects of the oil industry which will bring the greatest benefits over the longest period. The region remains confident that the benefits of the oil and gas industries will consolidate and continue well into the next century. Provided central and local government administrative and fiscal policies remain adequately coordinated, the worst effects of both rapid economic growth and national economic recession can be contained and largely diverted, thus providing an economic cushion for the nation during the 1980s, when national circumstances may otherwise be uncertain.

BEFORE NORTH SEA OIL

Before the discovery of oil in the northern North Sea, the region was recovering from the post-war period during which it had suffered the highest level of emigration of any planning region in the UK.[2] Mechanisation and farm amalgamation in agriculture, and over-capacity and low investment in the fishing industry had greatly reduced employment levels.[1] In the late 1960s improvements in agricultural management, increased fish prices and government investment in the fishing fleet began to stabilise employment. At the same time, as a result of government-initiated economic studies, north-east Scotland was identified as a development area which, combined with the introduction of regional employment premiums, encouraged further expansion of the small base of modern manufacturing industries. The combined effect was general stability of employment and population from 1968 until immediately before oil (1971), thoroughly justifying the combination of coordinated central government assistance and local government initiative.

During this period, most existing development plans had become inadequate and obsolete, partly because they were prepared during a period of severe economic decline. Many public services, particularly water extraction and supply, required modernisation, expansion and coordination under a single authority. Administrative boundaries, particularly of Aberdeen City, were drawn too tightly thus restricting the supply of land for development and complicating relationships between adjoining local authorities, particularly where land and infra-structure demands generated by the City were required to be met by the adjoining rural counties. During 1971–75, prior to reorganisation of local government in Scotland (May 1975), these problems were largely resolved through joint coordination of planning policies by the City and four north-east counties of the region. Additional financial support and the introduction of oil-related rate support grants encouraged and greatly facilitated local acceptance of the oil and gas industries in the early stages of their development and eased constraints on the expansion of Aberdeen and Peterhead. Today it is commonplace to hear 'the oil industry has been good for Aberdeen and Aberdeen to the oil industry'.

THE FIRST DECADE OF OIL-RELATED DEVELOPMENT

The search for hydrocarbons spread northwards during the 1960s, follow-ing gas discoveries in the Dutch sector and southern North Sea. The first commercial oilfield in the northern North Sea was discovered in 1969. The oil-majors were quick to realise that Aberdeen, the largest community in the north with an established harbor, airport, container railhead and scientific research facilities, was ideally located to service exploration and construction of offshore oil and gas development.[4,5] Aberdeen's harbor and airport came under increasing pressure to supply berthing and quayside facilities and improve and extend landing, hangar and personnel handling services. Since 1970 these improvements have continued, the harbor now being non-tidal and usable for 24 hours a day, readily accomodating both the North Sea fishing fleet and normal commercial trade alongside oil company supply bases and berths for supply boats. The numbers of these boats handled rose from nil to over 4000 per year by the mid 1970s, when offshore exploration and development was at its peak. Airport runway and passenger handling facilities have been extensively modernised and now handle almost 1.4×10^6 passengers annually, compared with about 150 000 in 1970, and helicopter operations, already the largest in Europe, continue to expand.[6]

Initial conflicts between oil companies and the fishing fleet centered on the damage incurred to fishing equipment from submarine debris and pipelines, and restriction to fishing grounds caused by the construction and operation of oil rigs and production platforms.[7] These have now been largely resolved through the formation of a fund and joint negotiating committee, responsible for settling claims and exchanging mutually valuable information.[8] A surge in the demand for pipe storage yards to service the offshore production phase was successfully met by local authorities who issued temporary planning permissions and restricted such development to environmentally less sensitive areas.

Apart from these changes, the casual traveller will not be startled by the visual impact of the oil industry, but should notice the continuing expansion of the City and surrounding towns, where major housing and industrial sites to the north, west and south of Aberdeen are being developed, and the rapid growth of new office and shopping facilities both in the City center and outer suburbs. During the 1970s, regional employment increased by nearly 30 000 jobs, most in oil-related industries; population increased by some 30 500 and over 30 000 new houses were built,[9] 55% by the public sector, where priority was given to the housing of incoming key workers. The simultaneous increase in the cost of private housing can be illustrated by the rise of 450% in the price of a three bedroom detached suburban house of traditional granite construction. These figures are only a guide since it remains difficult to accurately assess the effects of oil on local house prices in view of the increase in house prices nationally. The amount of serviced land for development is another indicator of impact, some 700 acres being taken up throughout the region since 1971, over 75% of which is located in the Aberdeen area. Perhaps one of the best indicators of economic expansion is average earnings, which have increased from 89% of the national average in 1968 to 106% by 1978.[10] Current proposals for major shopping and office re-development in central Aberdeen, and exhibition, recreation and leisure facilities reflect this increase in personal incomes. Further difficulties are experienced by non-oil industries in obtaining labor at reasonable wage levels and by public authorities in retaining or replacing administrative, clerical and semi-skilled staffs. Private home buyers, changing houses or entering the market for the first time, and groups on fixed incomes find the rapid increase in the cost of living an ever-growing frustration.

Perhaps the visual impact of oil-related development in the region is greatest in Buchan, at Peterhead, where development of the British Gas Corporation (BGC) St Fergus gas terminal, together with those of Total and Shell Expro, has been successfully integrated into the surrounding environment. The terminals mark the landfall of submarine gas pipelines, two from the Frigg and one from the Brent offshore fields. Three methane 36 inch gas pipelines and a fourth 42 inch pipeline, now under construction, link St Fergus to the recently extended national gas grid. Methane is separated from the 'wet' gases from the Brent field at St Fergus, from which NGL will be transported by undergoround pipeline to the Shell Expro NGL/ethane cracker now being developed at Mossmorran in Fife (see Taylor, these proceedings). Oil from the Forties (BP) field landfalls at Whinnyfold near Cruden Bay and is transported by pipeline for processing at Grangemouth in central Scotland. Pipeline routes were carefully studied to allow safe expansion of existing communities or new developments. At Peterhead, the Harbour of Refuge has been developed with the assistance of the Scottish Office and now contains two oil service bases and a new jetty to accommodate both the oil supplies to the new Boddam electricity generating station and the transport of gas condensates from St Fergus. The harbor is sufficiently large to accommodate the servicing of six–eight oil rigs per annum, with capacity for expansion. Industrial and housing areas in Peterhead and surrounding villages have been expanded with few adverse social effects, and up to 1600 temporary workers have been accommodated in labor camps during construction of the major civil engineering and pipeline projects.

Economic and Environmental Planning

Uncertainty has been a major problem for local and central government planners seeking to manage and control such large-scale development. Particular difficulty has been experienced in obtaining reliable advance information on the scale and rate of offshore and associated oil servicing activity, in monitoring the supply of land and infrastructural services and facilities available in development areas, and in rapid formation of land-use policies to guide development control and capital expenditure budgets.

To overcome these problems the Regional Council developed a system of integrated economic forecasts which are reviewed and updated annually.[11] These provide a basis for the preparation and coordination of local authority and government agency policies, and proposals by private developers. Economic forecasts are controlled by trends in the rate of land development at small settlement scale, and provide a basis for school roll and other secondary forecasts. Whilst the forecasting system has been instrumental in establishing priorities for local authority capital expenditure and ensuring adequate phasing of site development and servicing, its major purpose is to provide a ready source of information on economic trends, as a basis for the corporate preparation of land-use policies.

The region, following its establishment in 1975, was also required to introduce a new system of structure and local plans, with responsibility for preparing them shared with five District Councils, including Aberdeen City. It order to advise Districts, who largely retained the responsibility for housing and development control, the Regional Council prepared environmental policies in which some 14 types of regionally significant environment were identified and safeguarded from development.[12] The Council concentrated on completing a strategic or structure plan for the main pressure area of Aberdeen and 20 miles around in order to quickly revise outdated development plans and provide additional statutory safeguards and guidance for developers and District planners.[10] In addition, the development of contingency planning techniques has enabled local authorities, the Scottish Office, the BGC and petrochemicals operators to identify those areas within Buchan with potential for further pipeline landfall, petrochemical processing and the construction of additional harbor distribution facilities. The contingency plan both simplifies statutory procedures and will accommodate developments in the oil and gas industries, some of which may now result from the proposed North Sea gas gathering pipeline.[13]

A number of major factors have to be considered together to explain the level of planning guidance provided by local government during the first phase of North Sea oil and gas development. First, local and central government have successfully coordinated their administrative and financial policies, with the Scottish Office supporting local decision-making through the allocation of additional finance and the acceptance of economic forecasts informed by local conditions. Secondly, a 5–7 year delay between the exploration and discovery of offshore oil and gas, and the build-up of locally resident personnel for production allowed the necessary planning procedures to be developed and the most economic and desirable infrastructure to be provided. Thirdly, the rapid development of economic information forecasting and contingency planning techniques aided and guided decision-making by developers, industrialists and public agencies. So far the main reasons why there is little visual impact or serious environmental damage are, first, that downstream oil-related development has not yet taken place on any large scale, owing to the nature of oil activity in the area, e.g. there is fortunately no rig and platform construction in the Grampian Region, and, secondly, the success of local authorities, with the ready cooperation of the Scottish Office and the oil and gas industries, in the environmental integration of major developments. However, circumstances may change rapidly.

FUTURE IMPACTS OF OIL INDUSTRY

The most commonly held view of the 1980s has tended to be geared to a 5 year horizon, at the end of which the increase in oil activity levels out and then tails off, leading to the oft repeated query 'what happens when oil goes?' There have been good economic reasons to support this scenario, largely because of limited planning horizons adopted until recently by oil companies. In comparison with the mid-1970s, over the past 2 years there has been a significant reduction in employment engaged in offshore construction and a tailing off in exploration activity.

However, a number of factors have led oil companies to reassess their prospects in the North Sea.[14] The real sterling price of oil has risen during the past year, after little increase from 1975 to 1978. Political instability has continued in traditional oil-exporting areas, notably Iran. The attitude of the present government has been encouraging particularly in the support given for the gas gathering pipeline. Reinvestment potential of the North Sea has been reassessed, following the return of sizeable profits from initial investments. By September this year, some 15 offshore oil- and gasfields (23 fixed platforms) were in production, a further 11 fields had development plans approved, and it is possible that some 50–60 fields could be in production by the end of this decade.[14] However, uncertainties include

(i) some of the additional oilfields may be small and of short-term economic viability;

(ii) government depletion policy may slow the rate of both exploration and production;

(iii) rate of exploration, if not production, may increase particularly if the western Atlantic province is included and more large fields are discovered.

Certainly the proposed gas gathering pipeline system will reduce offshore flaring, boost gas industry supplies by a quarter in the late 1980s and increase the potential for additional petrochemicals processing locally and elsewhere in the UK.

Also pipe laying activity will increase in the early 1980s, with the offshore gas gathering pipeline, the NGL pipelines from St Fergus to Mossmorran and possibly to Nigg, and possibly a fifth BGC line to service the national grid. The proposed new gas terminal and possibly an SNG plant (BGC) will be completed at St Fergus and the recent announcement by Occidental to construct a polyethylene plant could introduce downstream petrochemicals processing to Buchan. In addition, the service bases at Peterhead Harbour may require expansion to meet the requirements of the oil-related sector.

Perhaps the most striking physical impact of the oil industry on the region during the 1970s was the development of prestige offices for the operating oil companies. With offshore operations being increasingly administered from Aberdeen rather than London, office accommodation is likely to double during the next 2–3 years, when the majority of oil companies will have offices in Aberdeen.[14]

Although growth rates will be within those expected (Aberdeen area structure plan), no flexibility will be available to District planners, which will require physical plans and policies to be reviewed, updated and rolled forward to 1991, much sooner (1982) than anticipated. The rural structure plan will require to take greater account of the implications for rural areas of increased job opportunities in urban centers and of the level of investment and support required to maintain the viability of small settlements. Major concerns center on the need for central government to take oil sector prospects into account if oil-related growth is to be maximised in the national interest, as local authorities in Grampian will be required to expend additional finance to facilitate it. Present cutbacks in Scottish Special Housing Association house building and the imminent withdrawal of regional aid imply a considerable shift of the financial burden from central to local government for meeting the demands of oil-related growth.

CONCLUSIONS

On balance, the advent of the North Sea oil and gas industry has been good for the Grampian Region, helping the area to recover from a long period of post-war stagnation during which its geographical location in the UK and in relation to overseas markets and its small manufacturing base provided an inadequate foundation for major economic expansion. Since 1969 the oil industry has been a catalyst to the local economy, helping it diversify and expand whilst providing an opportunity for upgrading, renewal and modernisation of infrastructural services and facilities.

During the 1970s the coordination of local and central government fiscal and administrative policies was largely successful. It supported local decision-making, with great emphasis on forecasting and monitoring economic change; it permitted development of plans to guide developers and to advance the scale and provision of services; it catered for uncertainty through contingency plans; sensitive environments were safeguarded and national agency activities were coordinated and monitored. Whilst the 1970s saw a level of development not experienced for some time previously in the north of Scotland, the 1980s are confidently expected to be a period of major consolidation, rather than boom, but with economic growth continuing at a high level. Despite these prospects, some circumstances differ from those of the early 1970s. Although considerable progress has been made in the renewal and upgrading of infrastructure, and in planning to accommodate additional development, it is clear that much of the flexibility built into land acquisition, zoning and servicing programs will be taken up during the 1980s and costly service thresholds will have to be crossed to meet anticipated growth levels. Furthermore, the time available for planning in the 1970s is no longer available, since many offshore projects are already at an advanced stage of development. Instead of 5–7 years, only 2 or 3 years may now be left before the major impact of the second phase of oil development is felt.

Finally, predicted growth in oil-related industries during the 1980s, whilst providing a further catalyst for continuing economic expansion, also provides a challenge to the concept of a balanced economy, which can be resolved only through continued coordination of central and local government policies and programs. A further challenge to both central government and local authorities is to continue contingency planning and to continue, with increased assistance from the private sector, previously successful programs of coordination and local support. Regular economic forecasts provide a basis for the adoption of common objectives by both private and public agencies. If the challenges outlined here are met successfully, local objectives successfully coordinated and local control maintained, the advantages of the second phase of oil-related development and consolidation can be maximised and an economic cushion provided for the UK during a decade when national economic circumstances may remain uncertain. Beyond 1991 forecasting becomes increasingly difficult in view of major decisions yet to be made. What can be said with some confidence, however, is that oil industry activity in the Grampian Region at the end of the century is unlikely to be less than it is now.

REFERENCES

1. Gaskin, M. North East Scotland Joint Planning Advisory Committee (NESJPAC). *1975 Regional report.*
2. Gaskin, M. (1969). *North east Scotland—a survey and proposals.* HMSO, Edinburgh, UK.
3. Gaskin, M. (1976). *Regional report 1976–81.* Grampian Regional Council, Aberdeen, UK.
4. By means of NESJPAC.
5. Royal Scottish Geographical Society (1973). *Scotland and oil.* Edinburgh, UK. Manners, I. (1979). *Planning for North Sea oil: the UK experience.* University of Texas, Austin, USA.
6. British Airports Authority (1970, 1979). *Annual reports.*
7. Oil Development Council for Scotland (1975). *North Sea oil and the environment.* Edinburgh, UK.
8. The Fisheries and Offshore Oil Consultative Group comprising representatives of fishing interests and the United Kingdom Offshore Operators Association.
9. Grampian Regional Council (1980). *Grampian population forecasts—summary of the 1980 update.* Aberdeen, UK.
10. Grampian Regional Council (1979). *Grampian Region (part) structure plan: Aberdeen area.* Aberdeen, UK.
11. Grampian Regional Council (1976–80). *Forecasts of population, employment and housing.* Aberdeen, UK.
12. Grampian Regional Council (1978). *Study of environmentally sensitive areas.* Aberdeen, UK.
13. Grampian Regional Council (1980). *Contingency plan for petrochemical industries.* Aberdeen, UK.
14. Grampian Regional Council (1980). *Oil related prospects in Grampian.* Aberdeen, UK.

Discussion

A. Jackson (Department of Social Anthropology, University of Edinburgh, UK; representing the Social Science Research Council, North Sea Oil Panel, UK). Mr Lapsley said the local authority had successfully integrated key workers with the resident population. In what context does he mean integration and how was it done? Similarly are there policies in Rogaland and Stavanger to enable foreign workers to be integrated with the community in which they find themselves? Lastly, are planners considering this problem?

H. A. Graeme Lapsley. The initial influx of key workers to the Orkneys was for construction of the terminal and accommodation was mostly on Flotta in a construction camp. Those who came with wives and families quickly found rented accommodation (there was surplus accommodation stemming from farm mechanisation). As they were temporary no particular provision for them was made by the local authority. Of the 288 permanent jobs, 230 have been filled by Orcadians, so there is not a large number of key personnel. Occidental has built a group of houses in the village of Finstown, but most permanent employees have bought houses individually throughout Orkney, have integrated well and all now play their part in community affairs. The local authority erected 24 houses on Flotta, most of which were initially occupied by terminal personnel, but many people, particularly those from the south, find the mainland of Orkney remote enough without committing themselves to the isolation of a small island and now the majority of tenants are islanders. The local authority does allocate one house in eight to technical (as opposed to lower management) workers at an economic rent when and where the houses become available, not in blocks, so that the incoming tenants are dispersed among the local community.

Since the time of Picts and Vikings Orkney has been no stranger to influxes of outsiders. World War II saw the whole of the home fleet in Scapa Flow plus four airfields and 60 000 troops. The locals have learnt to cope with such situations and turn them to their advantage.

K. E. Egeland. In the Stavanger area integration could have been better. The government has had fair success in stimulating Norwegian supplies to the oil industry but too little attention has been paid to the local economy. A clearer policy is needed to develop oil-related activities as a new market for existing companies, particularly in relation to various hindrances local companies encounter. For example, local industry has operated on a personal basis, largely of trust, and the system of big contracts, legal entanglements and so on is an anathema to them.

The real social impacts can probably be seen only in the long-term. However, the incoming workers in Stavanger are not the migratory workers at the bottom of the social scale but bring families and are at the top of the scale. The ghetto situation does not therefore generally arise. Stavanger has a large shipyard so for limited periods there may be large numbers of workers in construction camps, which do fluctuate in numbers. Although Americans tend to form their own communities, in general few problems of integration have been encountered.

T. F. Sprott. Not enough planning has been done on this subject although some work was done by the Scottish North Sea Oil Panel (see the General Discussion, these proceedings). However, there are a number of relevant factors in the Grampian Region).

The period of peak immigration (mid-1970s) saw a net inflow of 4500–5000 people per year, mainly to the Aberdeen or Peterhead area. In the face of considerable speculation (three new towns were variously proposed for the Aberdeen area) planning priority was given to locating new industrial or residential development as extensions of existing communities because

(i) infrastructure which was available could be extended more quickly and efficiently than construction of entirely new services;

(ii) newcomers could enter existing communities rather than starting from scratch in the middle of the countryside.

The results were mixed. Some communities grew far too quickly, some integrated better than others, and so on.

Increasingly, Grampian people are employed in oil-related industries and this itself leads to contact with the outsiders. Also there is a general movement of people from the inner city to suburban areas where the newcomers are also to be found, so new housing areas are mixed. However, community, social and recreational facilities have not always developed as rapidly as desirable in the new communities because of shortage of money and all that goes with it.

As it is largely oil-related personnel who suffer the shortcomings as well as the Grampian Region country people moving in it is in the interests of both the oil industry and local government to solve the problem. Although a company spending up to £2 million on an internal recreation center next to their office is to some degree justifiable there is a broader perspective. The possibilities of joint programs are now being discussed with the oil companies, in the hope of devising more socially equitable distribution of community facilities.

Anon. How does British Gas supply domestic consumers in Kirkwall (Orkney)? Does Occidental make concessions to its workforce because of the dual economy practised by many in the Orkneys?

H. A. Graeme Lapsley. British Gas's original intention was to supply gas to the Electricity Board in the Orkneys. However, plans were eventually smothered by bureaucratic argument. Occidental might comment further.

Flotta did have some locals employed part-time by Occidental but most are now full-time shift workers in a highly technical terminal, as running a croft is not so profitable and thought not as attractive although certain grants are available to crofters and shift work does enable a croft to be run, albeit with difficulty. This may be in part because Occidental is considered to be generous to its workforce.

J. H. Loosemore (BP Petroleum Development Ltd, Aberdeen, UK). In Aberdeen and the Grampian Region integration of the incoming workforce has been generally good. However, I feel minimum control is desirable and economic forces will largely exercise the control needed.

For major companies and local authorities to enter on joint ventures for recreational services, etc., is a good idea. Apart from during platform construction, though, the oil industry itself employs a fairly small workforce. The support industries, not the oil companies, require the greater part of the workforce and office space.

T. F. Sprott. Neither District nor Regional Councils wish to direct people where to live. Indeed legislation is now in process to remove the residential qualification from public housing lists, which should permit much greater mobility, particularly for people who seek local authority tenancies.

F. H. Mann (author). Mr Sprott foresaw the planning period fall from 5–6 years to 2 or 3 years, but that is too short a time to get any North Sea project started.

T. F. Sprott. The period of approximately 5 years between approval of oilfield development plans for the larger fields and their first production of oil is likely to fall for the increased number of small fields, due largely to (a) a reduction in offshore hook-up time and (b) the fact that some of the previously more marginal fields are already well advanced in oil company development plans.

Section IV

HAZARDS AND CONTROL

Chairmen

Papers 19 and 20
Hal Moggeridge Esq., President, The Landscape Institute, UK

Papers 21 and 22
Professor G. S. G. Beveridge, FRSE, Department of Chemical and
Process Engineering, University of Strathclyde, UK

Papers 23 and 24
William J. Cairns Esq., Senior Partner, W. J. Cairns and Partners
(Environmental Consultants), Edinbugh, UK

19

Environmental Control of Petroleum-related Industry on the Gulf of Mexico Coast

A. HARRISON

US Environmental Protection Agency, Dallas, Texas, USA

ABSTRACT

Transportation terminals and process facilities for production of oil from major new offshore sources are, for economic reasons, located in coastal zones. Coastal zones also have large population centers, prime agricultural land, beaches and vital wetlands. To protect health of residents and sensitive estuarine ecosystems, criteria for siting and controlling new, existing and expanding industry in the USA is through a complex system of local, state, regional and federal interests. The United States Environmental Protection Agency particularly covers air and water quality and management of hazardous wastes and toxic substances. Despite regulations many health and environmental problems continue to arise unmitigated. Generally in the USA, land-use planning programs are rejected in favor of corporate management and planning. However, where local support is strong, industrial siting can extend to secondary impacts such as community development and socio-economic benefits. American experience supports development of international agreements on pollution control measures and the sharing of knowledge among nations.

INTRODUCTION

Regional administrators of the US Environmental Protection Agency (EPA) are responsible for enforcing, fairly and firmly, the most stringent and thorough environmental legislation ever passed in the USA. Regions

211

are large, e.g. Region VI (the author's) includes Texas, Louisiana, Arkansas, Oklahoma and New Mexico, and administrators are deeply and daily involved in matters of vital importance to our existence on this earth.

Many conflicts are also involved. Often forces favoring environmental protection to the fullest possible extent are pitted against those who favor all-out energy development. Compromise is often the desired though elusive solution. These same conflicts often dominate elsewhere in the world, and the correct course of action seems to be the same also.

Ex-President Carter of the United States called the search for energy independence the moral equivalent of war. In recent years the USA has set about finding its way out of a crippling dilemma of consumption and supply. Yet we all know that this is not and cannot be all-out war. The search for energy cannot ignore the imminent danger to the quality of our environment unless, in another decade and already having suffered irreplaceable losses, we want to fight an environmental war. If we do, that war will be one much harder to win than the war against energy shortage. Rather, it must be recognised that energy and the environment are parts of the same war, with each perhaps the hero of his own side and the villain to the other (see Fischer, these proceedings). It is a war that neither side must be allowed to win at the expense of the other.

Sufficient energy and a clean environment are necessary for survival on this planet. We must have both and if either is allowed to destroy the other, then man himself will be destroyed. So it is the responsibility of all, whether in the production of crude oil or the protection of our most elemental of natural resources, air, earth and water, to function as aggressively and as competently as possible, but always in the knowledge that the goal is not unconditional surrender, but a negotiated peace.

Instead of all-out victory, we must apply ourselves to the slow, difficult, and often distasteful objective of compromise wherever compromise is necessary to the greater good. The greater good is survival. Compromise is quite a different thing from capitulation. Capitulation is always done from weakness. Compromises are made from strength.

In striving to restore national security and economic and even spiritual health, the USA is trying to reduce independence on foreign energy sources. At the same time, the USA is morally and legally committed to a long-neglected clean-up of our environment. So now in the USA we are struggling to find the common ground upon which our future will be built. The situation is critical, not just in the USA but all over the world. In 1977 President Carter asked all federal agencies to take stock of predicted changes in the world population, natural resources and the environment to the end of this century. The predictions are based on a continuation of present world policy. The results were made public this summer as the Global 2000 report and will serve as the foundation for much long-range planning. The results are disturbing.

Already the pressures of population on our resources are tremendous. Already these pressures deny many millions the basic necessities of food, shelter, health and jobs, with little hope for improvement. The pressures will increase, and more and more determine the quality of human life. Simultaneously, the ability of the earth's physical and biological systems to provide resources for human needs is steadily eroding. If present trends are allowed to continue, the result will be a steady degradation and impoverishment of the earth's natural resources and, soon after, of its inhabitants.

It is important to note that the impacts upon our oceans will be felt most significantly in the coastal zones. An important part of this will result from petroleum exploration and development and related industries, as they dredge and fill coastal marshes and estuaries for the installation of new transportation and industrial facilities, as they spill more petroleum products and toxic wastes in the course of going about their business. This is a worldwide problem and it is only with an era of unprecedented global cooperation and commitment that this planet will escape the deadly effects of its own negligence, past and present.

DEVELOPMENT OF OFFSHORE OIL SUPPLIES

The upper and western coasts of the Gulf of Mexico is the most active area of offshore petroleum and natural gas development in the USA. There is similar interest on the Atlantic seabord and the Alaskan coast, but those fields are still in the exploratory stages with many questions and many years between development and production. The coastlines of Texas and Louisiana are the best examples in the USA of long-term offshore drilling and its results: the establishment of a large and thriving support industry; the growth of related industries to a high density; the impacts of this growth on the environment both directly and indirectly; and how we as citizens and regulators have chosen to deal with them. It is obvious from the outset that there is a great deal of difference between this area and the North Sea, even if the obvious physical differences, in water temperature, depth and placidity, land topography and population density are ignored. The experience of oil production in each location remains vastly different.

Development of the Gulf has been a long process, which began in the 1930s and ambled along largely undisturbed until the last decade or two when concern for the environment began to grow and laws were passed and regulations instituted. Development of the North Sea fields has come largely within the last two decades. The activities have been much more similar to those of a boom town than to a long steady development. The Gulf development came after a large land-related support complex was already in place along the coast, while the Scottish coast was largely devoid of major development. So it is difficult to look at one area and predict what lies in store for the other, and even more difficult to decide what to do about it. But one can gain insight from an understanding of these differences.

When popular concern for the environment was spawned, the USA was presented with an array of established pollution. Pollution was well developed and had secondary and tertiary effects that ran deep into the fiber of society. Furthermore because of this and because of the national attitude toward land-use planning, we have only in the past half-decade become seriously involved in industrial siting. Parenthetically, there are a couple of constant factors in all of this. One is that in many cases the same companies that developed the Gulf are at work in the North Sea. Another, related, is that no matter who is at work or where they are working, there are always consequences of one kind or another, to one degree or another.

DANGERS OF DEVELOPMENT ON NON-PETROLEUM ENERGY RESOURCES

While Scotland's oil development is relatively fast, it is being mirrored in the south-western United States in development of energy resources other than petroleum. The potential in these areas is awesome and promises very much the same boom prospect as in Scotland. In Texas alone, 12.2×10^9 tons of lignite coal underlie the land in a broad band running south-west to north-east. It has been estimated that half of that will be needed by the year 2000 just to fuel the power generators projected for the region. The impact of such extensive mining, not just upon the earth that must be hauled up but upon fields and streams, the loss of farm and range productivity and upon the many hundreds of rural communities in its path, is frightening.

A further example is in New Mexico, where half the uranium reserves currently economically accessible in the United States lie in an area 90 miles by 20 miles. At present, there are 38 mines and five mills active in that area, with 33 more mines proposed as those are exhausted.

It is true of most mining, and even more true of mining the rare elements, that comparatively little of the desired element is found even in the highest grade ores. Many tons of rock are simply thrown away as tailings. Uranium is no different. Every mill in New Mexico already is flanked by extensive tailings ponds and piles. Uranium tailings are not just unsightly accumulations of rock and dust. They contain potentially serious concentrations of selenium, molybdenum, zinc, vanadium and sulfuric acid, in addition to radium and other radioactive compounds not useful in the production of nuclear fuel, yet still quite long-lived and potentially harmful to life. It is estimated that by 2000 A.D., some 3×10^8 tons of uranium tailings will have been piled across the San Juan Basin of New Mexico, potentially not only affecting the forage of grazing livestock but damaging the quality of already scarce groundwater supplies.

SOUND IMPACT OF RAPID DEVELOPMENT

Easily the largest boom town in the region is Houston, the Aberdeen of the south. Houston is one of the largest and fastest growing metropolitan areas in the country. This is not entirely due to petroleum development. The city is an international port, by air and sea, and it enjoys an excellent industrial and financial climate. But oil production and processing and the fabrication of chemicals and compounds from petroleum is at its heart. Because of this and because it has been in place for decades and not just years, Houston is an excellent example of the extensive effects of industrial development upon a society. One effect is that Houston has the worst air pollution in the region and one of the worst in the entire United States. This also is not only directly due to industry. One of the biggest contributors to air pollution in Houston is the private automobile, a source of pollution that grows with the city. The Houston metropolitan area covers 6285 square miles, but does not have a mass-transportation system that works.

The Houston ship channel is another example of secondary effects. Dug during the early part of this century, the channel stretches less than 20 miles from Houston's east side to Galveston Bay and the Gulf of Mexico. Its banks are crowded with industry, much of it petrochemical plants and refineries. Until a few years ago, the channel was one of the filthiest stretches of water in the world. It threatened to destroy the marine environment of Galveston Bay. It was so polluted by petroleum and petroleum products that a welder once set it afire with his torch. Today, stringent water quality laws and regulations and an outstanding effort by industry have made the channel much cleaner. Shrimp and dolphin have been found in its lower reaches, but the channel still experiences an occasional fish kill, through runoff from the city and its suburbs, and from the discharge of poorly treated sewage into the water. The city is growing so rapidly that it cannot build treatment plants fast enough to handle its own waste problem, in spite of intensive efforts by the City and the Federal Government (through the Construction Grants Program).

PLANNING AND CONTROL OF INDUSTRY, AND ENVIRONMENTAL LAW

The siting of petrochemical plants and other industrial facilities in the USA has not been a separate function of environmental law, partly because much polluting industry was already in place when new environmental laws came into effect in the early 1970s. The problem of controlling existing pollution was more important then than the siting of new plants. Also new sites were generally covered by the same regulations that applied to existing plants. In addition there is a general reluctance of the public and business in the United States to accept government management 'before the fact', a sentiment tied closely to the provisions for individual rights in the constitution, and formulated in years when our huge land area and our comparatively sparse population permitted it.

Although it is no longer tenable, that attitude is still the basis for much of the opposition to federal environmental regulation in the USA. It is changing slowly, however, with new laws and new amendments to old laws. It is being recognised that siting is a thing that requires attention.

Much of the change has come in the past 3 years and the new body of law that has emerged implies tremendous effects on the future siting of plants and the expansion of existing facilities. The combined effect of these regulations could well be the transfer of control over the questions of when, where and what types of industrial development will occur in certain areas of the country. Industrial corporations will retain the ultimate decision of 'yes' or 'no', but many of the remaining questions will likely not be answered without the review of government at the local, state or federal levels.

In general, environmental regulation in the USA is based on a system of restrictive permits and overlapping review carried out by the various agencies of government that are concerned with natural resources and environment quality. There are, for example, nine organisations that issue and review federal permits for activities connected with oil and gas development. Activities such as offshore mooring, platform fabrication, pipeline laying, and refinery construction are potentially subject to either permitting or review by at least eight different programs and in some cases eleven.

An enhancement to the system is the existence of the National Environmental Policy Act (NEPA) and the well-known environmental impact statement which it fathered. The purpose of those statements is to provide, in the case of projects expected to significantly affect environmental quality, a thorough study of the project and its potential impacts. The statements have to be detailed, comprehensive and readable. When they fulfil these three requirements they can be effective in aiding decisions made by the agencies involved and by the public.

All permits for dredging and filling issued by the US Corps of Engineers are subject to NEPA, as are new sources under the Clean Water Act, activities under the Coastal Zone Management Act and the projects included under the Resource Conservation and Recovery Act. In addition, more than one-third of all states now require environmental impact statements comparable to those created at the national level by NEPA.

Environmental law in the USA is generally organised according to the natural resource being polluted. The Clean Air Act provides for setting up national ambient air quality standards which specify maximum concentrations of pollutants legally permissible anywhere in the USA. The national standards cover major air pollutants (sulfur dioxide, particulate matter, carbon monoxide, ozone, nitrogen dioxide and lead) and the Act requires the attainment of these standards by certain deadlines. Attainment is achieved by limiting pollutant emissions from industry and from non-point and mobile sources. This is to be done with the cooperation of each State.

The States are charged with developing a State implementation plan which includes specific proposals for attainment and accompanying deadlines. The State plans are reviewed by EPA. They are regularly revised as progress is made and the States take more and more responsibility for the implementation of the programs under the Clean Air Act.

It is the general intention of many of our laws that EPA stop implementing specific regulations and operate a support system providing assistance and resources, review environmental effort and set standards for the States to achieve. Currently these standards are being attained through specific emissions requirements which are becoming increasingly stringent as technology becomes more and more effective and economically feasible.

In air quality, the siting of new plants is chiefly affected by the existence of special regulations dealing with areas of the country that do not meet national ambient air quality standards, and in areas that do comply by the Prevention of Significant Deterioration Program (PSD, a result of the 1977 amendments to the Clean Air Act).

PSD requires preconstruction approval of new major plants proposed for areas that meet the national standards. The law provides for these areas of attainment to be divided into three classes, each with a different limitation on air quality deterioration. Class I includes pristine areas where minimal air quality deterioration is allowed. Class II includes areas of moderate air quality deterioration and Class III includes areas of greater deterioration and thus allows for more industrialisation. At present, no area in the country is designated Class III. Any new plants permitted in these areas are also subject to the installation of best available control technology. The States are responsible for the classification of attainment areas and the States can alter the classifications. But the procedure is difficult, including public review and comment. Public hearings and a detailed analysis similar to an environmental impact statement are required for the affected areas. If an area is to be reclassified to Class III, it must further be approved by the Chief Executive of the State and the State legislature. That done, it must also be approved by local governments representing a majority of the residents in the area.

Restrictions on new plants in non-attainment areas are more stringent. They operate on the simple rule that new sources must not add to the emissions load already present in the area. They require the installation of the latest control technology. New plants must accept stringent controls and existing emissions in the area must be reduced to more than compensate for the emissions of the new plant before the new plant can be built. This is the heart of emission offsets, a policy that allows growth consistent with improving air quality.

In these areas specifically, EPA is trying to get industry more involved in self regulation. One such attempt is the 'Bubble Policy' which, instead of restricting each emission source at a plant, restricts emissions from the entire plant, or even a complex of plants, as if under one dome or bubble.

Emissions coming out of that bubble must not exceed certain levels. Plant operators are given the choice of finding the most efficient and effective way of holding emissions to those levels. A program that is still in the talking stage is one that would involve the accumulation and possibly even the trade or sale of emission credits. Under this program a company could voluntarily control its emissions to a greater degree than is actually required and receive 'credits' for a certain amount of emissions. Then it could either sell those credits to a company in need of more emission space or spend them at a later date on its own plant expansion.

The Clean Water Act has a simple premise. It prohibits any discharge of pollutants into public waters without a permit imposing control requirements. This applies whether the discharge is existing or from a new source. EPA has consequently issued new source performance standards for a great variety of industrial categories, which define the specific levels of pollution control required of each plant according to the processes underway within that plant. Furthermore, an environmental impact statement is required for new sources if performance standards exist for that source, and if the source would have significant effect on the quality of the human environment. The impact statement in any case must be submitted and acted upon before construction is allowed to begin.

The regulations of the Clean Water Act affect many aspects of the petroleum and petrochemical industry, including drilling rigs, refineries and petrochemical plants. The greatest effort so far has not been in siting but in regulating wastewater discharges. Until recently EPA had issued no permits for oil and gas activities in the Gulf of Mexico, despite having had the authority to do so for some time. Regulation in this area was slow to start with because EPA lacked the resources to issue permits for every one of the approximately 2500 exploration, drilling and production platforms in the western Gulf.

Two developments have changed that policy. First, EPA has increasing suspicion that substances found in drilling fluids may be more hazardous to marine life than was once thought. Secondly, EPA has been granted the use of the 'general permit', a regulatory mechanism that allows us to permit and restrict all oil and gas activities in the Gulf without having to write a permit for each individual facility. Three proposed general permits would cover leases in the territorial waters of Texas and Louisiana and in the federally owned areas of the outer continental shelf. They are now open for public comment prior to issuance.

All EPA permits must undergo a lengthy period of public review and comment. Following that, the comment collected is reviewed and if numerous or substantive enough it can delay issuance or even substantially alter the requirements of a permit.

As proposed, these general permits would be of 2 years duration. They would prohibit the discharge of halogenated phenolic compounds, oil muds and detergents used during tank cleaning operations. They would also require regular monitoring of the flow rates of produced water, check drainage, sanitary wastes, drilling fluids and cuttings, and the concentrations of oil, grease and chlorine in the discharges.

Areas not included in these permits are several potentially productive or unique biological communities in the Gulf leases. Special permits are being written and proposed for the areas. The Flower Gardens banks in the northern Gulf are a current example of this. They are the northernmost living coral formations to be found in the Gulf. They rise to within 10 m of the surface and they are deemed unique and worthy of protection.

Several efforts underway to provide that protection are an example of the overlapping effect of federal regulation and some of the conflict referred to above. Some of the conflict comes from within government itself. EPA, the Department of the Interior and the Department of Commerce all have an interest in the lease areas and in the Flower Gardens, but they disagree somewhat on how protection should be achieved.

EPA proposes to issue 12 permits under the National Pollution Discharge Elimination System and supports Department of Commerce efforts to protect the banks further by establishing a marine sanctuary. The EPA permits prohibit discharges in areas of the banks that are less than 100 m deep and provide a lesser degree of protection in the areas surrounding the 100 m isobath.

The Department of the Interior favors an 85 m isobath and in the surrounding area requires monitoring and the shunting of drilling fluids to within 10 m of the bottom. It also opposes the establishment of the sanctuary as unnecessary. It is fairly clear that what is also at issue here is which agency has the dominant authority in the area of offshore leases. This conflict can also be seen to some degree as one involving production versus protection and is thrown into even better focus by the addition of the various citizen-supported environmental groups to the fray. The National Resources Defense Council, for example, a group already moving to contest these permits in court, argues that they are too lenient and that EPA cannot issue them without a greater knowledge of the effects of drilling fluids on the marine environment. There is, of course, a compromise somewhere, and we are in the process of finding it.

Two programs that hold a great actual and potential effect on sitting in coastal areas are the President's Preservation of Wetlands Executive Order 1977 and Congress's Coastal Zone Management Act 1972. Both are administered by agencies other than EPA, but are subject to EPA review and comment. It is EPA's responsibility during review to supply a checklist of environmental criteria and then to judge how well they are being met.

Currently, the weight of Presidential support is against construction in wetlands unless there is no practical alternative to such construction and all effort is made to reduce the potential harm. The result of the executive order has been that it is now more difficult to obtain a permit to fill wetlands. According to the law, if a facility's construction plans involve dredging and filling of wetlands, a permit must first be obtained from the Corps of Engineers and undergo an EPA review.

In hopes of stimulating land-use planning and controls in coastal areas, congress passed the Coastal Management Act and set aside for the States substantial planning and operating funds. The effectiveness of the Act has evolved slowly, but it retains the potential to become the most important control of siting in coastal areas, those areas most significantly affected by offshore drilling. The program will shift control of coastal development from local government towards a more centralised state organisation. The program provides for identification of the boundaries of the coastal zone in each State and defines what shall constitute permissible land- and water-use within that zone. It also requires a description of exactly how the State proposes to enforce its plan; all of this is subject to EPA review.

The Resource Conservation and Recovery Act 1976 is symbolic of the effects that industry can have on entire communities, even whole countries, and of our continued ignorance of the effects of many new chemicals and compounds being produced today. It is no mystery or surprise that those new products, may of which are very beneficial, are the result of the age of chemistry, or of petrochemistry, that has exploded upon us since World War II. Many result directly from the cracking of hydrocarbons out of petroleum and they have come so rapidly that we do not know all we need to about their effects, good or bad.

One thing we have learned, however, is that many of the hazardous wastes generated in the USA come directly or indirectly from the petro-chemical industry. The Resource Conservation and Recovery Act attempts to regulate the disposal of solid wastes while paying special attention to hazardous wastes. It initiated a new permit program for facilities that dispose of hazardous wastes, and it created a 'cradle-to-grave' system to monitor final disposal. Under this law, wastes are judged hazardous according to their ignitability, corrosiveness, reactivity and toxicity. In addition, more than 100 specific industrial waste streams are listed because they exhibit one or more hazardous characteristics.

The law provides for development of solid waste plans by the States on environmentally sound guidelines and for development of criteria for sanitary landfill and for inventory of facilities that do not meet the criteria for sanitary landfill. The Resource Conservation and Recovery Act was not the legacy of Love Canal, the chemical dumping ground in a residential area of upstate New York, but much of the activity under the auspices of this law since the Love Canal discovery has been. For years Love Canal was used by the Hooker Chemical Company and there is currently underway an investigation into reports that the USA Department of Defense also used the site to dispose of extremely toxic wastes. The dump was covered over, presented to the local school district and turned into a playground. It was an extremely dangerous playground. Houses were built on it. The discovery of extremely toxic wastes and their effects of the populace of Niagara Falls has caused a furore in Congress and across the country.

A little over a year ago EPA, as a result of a national survey, listed 32254 sites where hazardous chemicals were known to have been stored or buried. Preliminary inspection determined that more than 800 of them were so poorly designed or managed that they posed 'significant imminent hazards' to public health: efforts are currently underway to bring those facilities under control. Production, handling and disposal of most hazardous wastes is still a mystery to us, hence the cradle-to-grave concept of regulation. It has been determined that these wastes are too difficult to control otherwise, and too dangerous not to control well. This problem, the severity of it, was unknown for a very long time. The petrochemical industry began in the 1940s and with the availability of cheap petroleum it has flourished ever since.

One of the greatest unknowns that remains with us is the long-term health effects of the chemicals and compounds produced by this industry. We are in the hands of the worst outbreak of cancer in the nation's history. Epidemic or not, 20% of all those who die by disease each year in the United States are killed by cancer. All the specific causes of cancer are not known, but scientists are wondering if at least part of the cause can be attributed to pollution. Considerable study is underway at present to try to determine whether this is true. Some areas of the country show much higher incidence of cancer than others. Scientists are trying to determine what is different about those areas.

CONCLUSIONS

Scientific investigation is the key to unravelling this problem, just as it has been the key behind most of our environmental legislation. Behind our laws lies scientific fact and behind our regulations the presumption that we would rather, when it comes to public health, enforce regulations that are too strict than too lenient. The more investigations we make, the better we are able to tie these long-term secondary effects of industrial development directly to the industries themselves. The same is true of effects that fall into the areas of social and economic difficulties. One only has to look at a typical boom town like Houston to recognise this. With the protection of coastal zones and the availability of substantial funds EPA hopes to better control those problems.

Progress is being made. One of the most serious problems in our system of regulation has been that of the regulators always being in a reactive position, arriving after the fact and with too little information and too much pressure. The environmental laws enacted since 1970, however, have forced corporations to change their planning methods. A company can no longer choose, buy and build on a site without first undergoing extensive review at federal, state and, increasingly, local levels of government. Corporations must now plan much farther in advance. They must incorporate environmental regulations into their own siting proposals. All levels of government have more time to study siting proposals and to determine their advisability.

Changes such as this in the way things are done, not just stop-gap measures of control, will have deep and long-lasting effects on pollution that will help get us through the present and into the future.

BIBLIOGRAPHY

Anon. (1980). *Refinery siting handbook*. Prepared for the US Department of Energy, Assistant Secretary for Resource Application, Office of Oil and Natural Gas, Resource Application, Contract No. DEA COI-79RA33001.

Brown, M. (1980). *Laying waste*. Pantheon Books, New York, USA.
Council on Environmental Quality and the Department of State (1980). *The global 2000 report to the President: entering the twenty first century*. 3 vols. US Government Printing Office. Ref. No. 0-274-484.
Quarles, J. Jr (1979). *Federal regulation of new industrial plants*. Box 998, Ben Franklin Station, Washington, D.C., USA.
US Environmental Protection Agency (1980). *Environmental outlook 1975–2000, Region VI*. Office of Strategic Assessment and Special Studies, Office of Research and Development, USEPA, Washington, D.C., USA.

20

The Shetland Islands and the Impact of Oil

J. M. FENWICK

British National Oil Corporation, Glasgow, UK

ABSTRACT

The Shetland Islands are small, remote and had only traditional industries before oil-related development. To optimise the effects of development of the Sullom Voe oil terminal on the Islands the local authority sought advice from consultants, initiated private legislation, prepared plans to establish where development would be allowed and to control its operation, and established management agencies in association with major oil companies. Development was eased by the presence of forceful individual public executives. Urban areas experienced great commercial and administrative expansion with associated multiplier effects. Airport, harbor, road and other infrastructure has been greatly improved. Some of the extensive building activity runs counter to traditional settlement patterns, labor has been drawn away from traditional industries, crime has increased and land and housing have become increasingly expensive. While the local authority has exercised effective control and there have been many benefits from development of the terminal, adverse effects may be felt more strongly when the construction phase is over.

INTRODUCTION

Shetland was, and still is, a small, remote and relatively unsophisticated community. Traditionally, the three main industries of fishing, knitwear and crofting were often represented in one family. In 1972 the population

of the Islands was about 17000; now it exceeds 22 000. The pattern of
settlement is rural, scattered and small-scale except in the main town of
Lerwick which accounts for about one-third of the total population.
Good farmland is rare and does not compare with the richer lands in
other parts of Britain. Conservation of cultivable land and improvement
by liming and fertilisation of other areas is essential for some degree of
self-support. The hills are peat covered, undulating, and all below 1500 ft.
The landscape is almost treeless, its beauty coming from the interplay of
land and sea.

THE APPROACH TO OIL-RELATED DEVELOPMENT

Shetland has been complimented on its handling of oil-related development. Four main groups of factors distinguished Shetland's approach
to that development: initial attitudes, planning and preparation, communication and responsibility, leadership and direction.

Initial Attitudes

In 1972 the British government had few guidelines in respect of oil
resources or developments from which affected areas could draw comfort
or inspiration. It was left to the affected communities to begin their own
planning and dialogue with the oil companies. May be, due to its relative
isolation, this was more noticeable in Shetland than elsewhere; perhaps,
in turn, the need for control became more sharply focused. Whatever
the reason, some areas fared better than others.

Shetland and Orkney could afford to take a strong line. In contrast to
other parts of Scotland where unemployment was rife, the Islands'
economy was healthy and based upon several industries (fishing, agriculture and knitwear) which had been tried and proven over centuries.[1] For
the first time for a century the 1960s actually saw an increase in population
rather than widespread emigration. This stability, new-found though
it was, was to Shetland's advantage. The islanders could say, hand on
heart, that Shetland did not need oil; if it was to come, it would come
on Shetland's terms.

Of course, that was a simplistic way of looking at it, albeit a good
standpoint for negotiation. Where there is oil, the oil companies have a
reasonable wish to exploit it and national governments a reasonable
wish to take benefit from their resources. Shetland had to work within

that context; it would scarcely have been realistic to play at King Canute. This philosophy is nowhere more clearly demonstrated than in one of the first policy statements made by the local authority.

'This Council, recognising that it may be in the national interest that Shetland be used for oil installations, and having sought to devise policies and to provide machinery which recognise the national interest while protecting those of the Shetland Community, will continue to have regard for the national interest but will give no encouragement to developments and will oppose proposals where these developments or proposals put Shetland at unnecessary risk or fail to provide available safeguards and will at no time put commercial or industrial interests before those of the Shetland community'.[1]

It was a statement which found a high level of acceptance throughout the Islands. Between those who feared that the advent of oil would be an unmitigated disaster and those who said that it would ensure jobs at home rather than on the vessels of the South Atlantic whaling fleet, there was a mutual concern that the Shetland community and environment should be protected.

Planning and Preparation

The first steps towards achieving this protection were taken in July 1972, the month which also heralded the discovery and size of the Brent field, a super-giant of 1.7×10^9 barrels of crude, 5.3×10^8 barrels of NGL and 3.5×10^{12} cubic feet of gas. To aid their planning, the Council hired some outside expertise. Experience was to show that consultants operated best where their role was technical and their remit specific. Both of those conditions were present; they were commissioned to advise on potential sites for major oil-related development. They were asked to consider marine and engineering aspects and they concluded that the most likely candidate was the Sullom Voe/Swarbacks Minn area.

The years 1972–74 have been described as the period during which Shetland bought its ticket to the dance; during which, in the words of one of Sir Kenneth Alexander's† predecessors, the oil companies were to arrive in Shetland to find the local forces already armed and waiting. They were attractive metaphors, but overlooked the plain hard work involved. For the benefit of government and of other areas which

† Chairman of the Highlands and Islands Development Board.

may face a similar situation, it may be useful to select three elements of that work.

First, between July 1972 and March 1973 the local authority prepared a development plan for the whole of Shetland.[1] It was a short, broad-brush interim plan—everything which is most enjoyable and irresistible to planners the world over—but it was of fundamental importance and it worked. Many of its policies are as valid today as they were when it was written. It established where development would be allowed and how it would be allowed to operate. In a climate of furious land speculation it directed all major development to one area, Sullom Voe; it allowed for lesser development in areas where limited job opportunities were desirable; it proposed that major installations would take place only on land under local authority ownership and only on the basis of joint use by the companies concerned. The oil companies, who may have had very diverse ideas and proposals, were thus asked to perform as one industry, in one place and with the Council continuously involved as landlords.

Secondly, in autumn 1972 private legislation was initiated which was eventually to result in an Act of Parliament, the Zetland County Council Act 1974. It was perhaps that element of work which has been of most interest to outside observers and which more than any other epitomised the Shetland approach. Perhaps what is not so widely appreciated is that the oil companies, who could have been in no doubt as to its implications for them, made no attempt to block its progress. There are many things about the oil industry, because of its magnitude and the very nature of the products it handles, which can cause fear in the minds of people living in its new territories. It has worldwide apparently ultra-sophisticated and continually developing technology which is often in direct contrast to the local, not very sophisticated, sometimes under-developed, communities which it affects. But surely an ideal situation begins to be reached when the industry's expertise and technology can be harnessed to work towards rather than against local aspirations. The assumption is that both parties act responsibly and in recognition of the other's objectives.

The powers conferred on the Shetland Islands Council by the Zetland County Council Act 1974 are certainly wide-ranging and demand a high degree of responsibility in being exercised. They include

   (i) powers of compulsory purchase over the land areas needed for major installations (revenues from land lease arrangements);

   (ii) powers to take on the duties of port and harbor authority in the Sullom Voe area, involving the local authority directly in port control, navigation, pollution clean-up, pilotage, tanker jetty construction and so on (revenue through the Ports and Harbours Agreement and harbor dues);

(iii) powers to control offshore works within 3 miles of Shetland, including dredging and pipe laying activities or, conceivably, the stationing of floating or submersible storage tanks or process barges;

(iv) powers to engage in commercial enterprise on behalf of the community, for example the Council's partnership with a major Clydeside company in providing towage services within the port (revenue from commercial interests);

(v) powers to establish a reserve fund out of revenues accruing from oil developments for use in social and community aid. Shetland knows that oil is a finite resource and sees 7 years of plenty being followed by 7 lean years; the reserve fund is a store of corn for the future.

People have said that no other local authority in Britain would have been given such powers, the implication being that, in 1974 at least, Shetland was so remote as to be eminently forgettable by Westminster and the rest of Britain. Certainly Shetland, with Orkney, had the advantage of being an identifiable community and a single local government unit. The fact is, however, that Shetland had a clear objective and arguments which were right and proper for the situation she faced. Parliament's assessment of those arguments was severe and thorough and Shetland's case was supported by a number of important national bodies.

The third element of preparatory steps taken in 1972–74 was completion of a detailed master plan for the Sullom Voe area in January–September 1973.[2] It assessed likely reserves of oil and gas both east and west of Shetland; it looked at the marine and landward features of Sullom Voe as a deep water terminal; it reviewed the possibilities of different types of development (e.g. oil terminal, refinery, LNG plant, transshipment port); it designated areas suitable for the likely facilities, making sure each could be accommodated; and it looked at where the new population might best be accommodated, selecting four villages for expansion. After a great many public meetings and considerable amendment it was approved by the Council in September 1974 without a single outstanding objection. In the same month, the Council approved (with provisos) the industry's own development plan for the terminal and the first of 1.3×10^7 tonnes of peat were removed from the site's construction camp.

Communication and Responsibility

In Shetland, effective dialogue between industry and community was an explicitly stated objective.[1] From the outset, the Council asked several of the majors involved in offshore exploration to join with them in an Oil Liaison Committee. Later the Sullom Voe Association (SVA Ltd), a non-profit-making company, was established as a partnership between the Shetland Islands Council (as A members) and the two major pipeline groups (as B members) for whom Shell and BP acted as spokesmen.

SVA Ltd is a unique arrangement not only for exchanging information but for supervising construction and operation of a huge development. It has five advisory groups with outside expertise, including universities, as appropriate. Two groups deal with oil spill procedures on land and offshore respectively; the others are responsible for technical, marine and environmental matters. The Sullom Voe Environmental Advisory Group, drawn from several of the organisations represented at this conference, produced good advice and an environmental impact assessment for the Sullom Voe Terminal.[3] The group was also an attempt to achieve a vehicle for reasoned discussion about environmental matters before conflicts arose. In practice, it has not proved fully effective, perhaps because it was not fully integrated with management structure. Perhaps the day may come when environmental considerations can be brought in from their fringe role to become a central function of decision-making.

Leadership and Direction

Shetland gained incalculable benefit from having Ian Clark in post at the right time. In addition, Shetland has been fortunate in having several political leaders of high quality.

There are strong parallels between the community impact in the Orkneys and in Shetland (Lapsley, these proceedings). Both island communities are sufficiently compact, robust and homogeneous to allow a high degree of both accountability and participation. The degree of community participation in the early days of planning in the Sullom Voe area was quite remarkable. For example, in the first three of 24 public meetings in 1973–74, Council representatives and their consultants were able to speak directly to literally everyone of voting age (and often below) who lived in those areas and was physically capable of attending. Those meetings began in a climate of misgiving which improved considerably once the community saw that their views were being genuinely considered and acted upon. That, in turn, meant that elected members and officials had confidence in, and commitment to, their cause. It was Winston Churchill who said[†] that he had 'never accepted what many people have kindly said, namely that I inspired the nation. It was the nation that had the lion heart. I had the luck to be called upon to give the roar'.

THE IMPACT OF DEVELOPMENTS

The impact of oil-related developments should be viewed against the background of three points. First, it is a huge development.[5] The site of the Sullom Voe crude oil terminal covers about 1000 acres plus an area reclaimed from the sea into which massive quantities of peat overburden have been poured. The peak construction workforce was about 6000, mostly non-Shetlanders, and housed in one of two construction camps or in accommodation ships. More than 30 oil companies are involved as participants in the Brent and Ninian pipeline system and this in itself has required complex management. The terminal contains facilities to receive, process, store and transport 1.5×10^6 barrels of oil per day, from 1982, with capacity to almost double the volume should a third pipeline be introduced.

Secondly, there are other important development areas. Lerwick has experienced great commercial and administrative expansion, growth in personal and professional/financial services, in building activity, in office development and industrial sectors such as plant hire, office supplies, distribution and warehousing. It has also become the center for supply base activities in the Islands. Some unusual craft have appeared in the harbor, not least the *Viking Piper,* BP's main lay-barge on the Ninian line, which appealed to the maritime interests of Shetlanders. Airport facilities have developed at Sumburgh and, to a lesser but growing extent, at Unst, the extreme southern and northern tips of the Island group respectively. At Sumburgh, total number of aircraft movements in 1969 was 1900, in 1977 39 000; passengers in 1969 totalled 26 000, in 1977 404 000, and in 1979 over 800 000; total employment in 1969 was 7 people, in 1977 171 people. This growth has brought new construction at the airport itself; an extended runway, two new terminal buildings, new

† At Westminster on his 80th birthday.

radar, fire-fighting, maintenance, office and hangar developments and, as in Sullom Voe and Lerwick, has subsidiary effects in a wide area around.

Thirdly, because of the Island group size and scattered nature there are still areas in Shetland where the impact of oil-related development has been marginal. The outer islands such as Foula, the fishing strongholds of Whalsay and Burra, the industry problems of which are not predominantly related to oil, and the vast wilderness and crofting areas of West Mainland are so far relatively unaffected. Pipe laying and dumping rubbish on the seabed, together with the nature and quality of effluent discharges from the terminal, have been sources of real concern to the fishermen of the Islands and it goes without saying that their communities, with every other part of Shetland, would be caught up in the consequences should the worst happen. The threat of a major pollution incident is undoubtedly their greatest fear. It hangs like a black cloud over the fishing industry and the great bird populations of the Islands, and will only leave, and perhaps not even then, when the last tanker leaves Sullom Voe (tanker traffic may continue in the western approaches and North Sea).

Among the other physical or 'environmental' impacts have been colossal growth in road, sea and air traffic, but the main problem has arisen with increasing numbers of heavy vehicles on roads not designed for them. Unlike pollution, however, there is at least a silver lining in this particular cloud in that when all is completed Shetland will have a much improved road system.

The logistics of peat disposal from road improvement schemes, not to mention the terminal itself, have also presented engineers with unfamiliar problems. The terminal site at Sullom Voe was, in environmental terms, an ideal place to develop. Very few people live anywhere near and there are no particularly important wildlife habitats, landscapes or agricultural areas in the vicinity. In fact, much of the surrounding area was in a state of near dereliction following its use as a wartime base. However, physical effects from such development are felt in a wider area, for example due to the need for quarrying activities.

Population growth has meant considerable building activity,[4] mostly group housing schemes which run counter to the traditional scattered settlement pattern. Several villages have been expanded to accommodate the incoming population.[5] Although the Council, aided by the industry, has put a lot of resources into improving community facilities in line with housing developments, incoming personnel may find life difficult in townships such as Brae. Brae is small with little choice of shops or entertainment and the long dark winters can be bleak indeed. For some, there will be compensations in the freedom of the countryside, the scenery, wildlife, the pleasures of sailing or fishing or mixing with easy-going resourceful people; others will not settle so easily.

Economic impacts have been considerable. The oil industry in Shetland has made labor expensive and has drawn it away from some of the traditional industries.[6] Many local people have become involved directly in short-term construction work; young people have forsaken college or university for the more immediately lucrative pursuit of driving a truck at Sullom Voe. No one can democratically stop it happening, but the task of retraining these youngsters for a long-term trade or profession will have to be faced and less than 100% success must be expected. Other jobs too are at risk. For every direct construction job, a multiplier effect creates another in sectors such as quarrying, plant hire or servicing. These too, once the construction peak is over, will disappear, and some are on the point of doing so now. Property values and house prices have leapt upwards. When the easy money goes, some of the mortgages are going to feel very like millstones. Small local firms have found it difficult to compete for the bigger, more specialised contracts. They have consequently lost labor or have been taken over by bigger outside interests with only transient commitment to Shetland, lasting precisely as long as large profits.

Socially, the twin pressures of increased population and increased wealth have combined to change attitudes and life-styles. As many other remote communities, and better than some, Shetland has been able to adapt to and absorb the technological trappings of the 20th century. Like many other communities, she has found that these things have made life easier. The process of rapid industrialisation is not so easy and, once community attitudes and expectations have changed, particularly in the young and active age-groups, history indicates that there is no turning back. Given the natural resources of the Islands, some new means to maintain the Sheltand community in the 21st century must be found, as important a job as the job done during 1972–74.[6]

The size and nature of the construction workforce, all short-term and residing in Shetland on a single-man basis, has caused problems but not as many or as serious as was first feared. They have been provided with excellent residential and social facilities and any misdemeanors have been quickly dealt with by the employing companies.

There has been an increase in crime, but offences have generally been minor. Nevertheless, in 1973 there were 12 policemen in Shetland and there are nearly 50 now. These problems may be reduced as the construction phase diminishes and the terminal settles into operational life.

CONCLUSION

There are probably as many interpretations of the impact of oil on Shetland as there are Shetlanders and oil men. Some have been unshakeable in their belief that it can mean nothing but harm; others have taken substantial material benefits from its presence. Some years ago, a revealing piece of research was carried out into the attitudes towards oil of two very different areas of Shetland, the Sullom Voe area and an area relatively untouched by developments, West Mainland. The people around Sullom Voe were generally more sympathetic to and less critical of oil developments than the people of West Mainland. Were the people of West Mainland expressing a fear of the unknown or was it that they could see the disruption without participating in the benefits?

REFERENCES

1. Anon. (1973). *Interim county development plan.* Zetland County Council, Lerwick, Shetland, UK.
2. Livesey & Henderson (1973) *Sullom Voe and Swarbacks Minn area: master development plan.* Zetland County Council, Lerwick, Shetland, UK.
3. Sullom Voe Environmental Advisory Group (1976). *Oil terminal at Sullom Voe environmental impact assessment.* Thuleprint Ltd, Sandwick, Shetland, UK.
4. Anon. (1979). *Shetland's oil era* (revised edition). Shetland Islands Council, Lerwick, Shetland, UK.
5. Anon. (1974). *Sullom Voe district plan.* Zetland County Council, Lerwick, Shetland, UK.
6. Anon. (1979). *Shetland structure plan.* Shetland Islands Council, Lerwick, Shetland, UK.

21

Hazard Analysis: What Can it Achieve?

D. H. SLATER,[a†] C. G. RAMSAY[b] and R. A. COX[a†]

aCremer & Warner, London, UK; bCremer & Warner Scotland, Aberdeen, UK

ABSTRACT

Process plant hazard analysis is developing to meet the need to include abnormal events in environmental impact studies and to build safety into plants rationally rather than by routine. Different forms of hazard analysis, each of which has an appropriate application, include risk assessment, hazard and operability study, quality control and, during normal plant operation, safety audits. Previous hazard analyses and the part played by hazard analysis in North Sea hydrocarbon-related development in Scotland are reviewed.

INTRODUCTION

This paper is largely directed at the validity of hazard analysis as a tool in decision-making. To fulfil this role hazard analysis needs to be accepted as a considered, tried and tested technique, not just a current technological placebo.

The perspective from which conclusions can be drawn on the evolution, nature and applicability of hazard analysis requires

   (i) review of the history of development of the technique;
  (ii) definition of the various present forms of hazard analysis;
 (iii) comment on the appropriate application of each.

† *Present address:* Technica Ltd, 11 John Street, London WC1N 2EB, UK.

Particular examples of onshore facilities associated with North Sea oil can be cited to put the use of hazard analysis in context, and as a starting point for considering future trends.

HISTORICAL PERSPECTIVE

Hazard analysis, although fashionable at the moment, is not a revolutionary new technique. It has evolved over several decades in which many lessons have been learned. The technique is still evolving and it is still perhaps too early to make definitive statements, but potentially it is of great significance to society because for the first time it enables us to weigh up risks in a rational way before we actually expose ourselves to them, instead of simply learning by bitter experience.

The historical development of risk assessment from an instinctive basis to an objective and rational one spans many centuries. For the caveman, taking on mammoths and sabre-toothed tigers involved entirely intuitive risk benefit analysis. There are documented examples from the 14th century of maritime insurance where traders, out of common prudence, sought to offset losses in natural disasters (acts of God). Since the Industrial Revolution man has had the power to affect the community in a similarly catastrophic fashion. Examples which show the impact of this power include the Tay Bridge disaster[1] and some of the more unpleasant pollution and industrial accidents.

The Swedish industrialist, Nobel, devoted considerable effort to development of explosives which could be used more safely and was demonstrably successful in correctly assessing the critical areas of risk. Particular historical landmarks, however, were the World Wars, especially the second, which demonstrated the power of applied science and technology and has been identified as the start of our modern, highly technological society (e.g. aerospace). Application of statistical techniques to allocation of spare parts and planning of maintenance schedules was an important concept which produced much benefit in the logistics of battle. This evolved into the techniques of systems analysis, which flourished in the post-war environment. World War II also spawned the nuclear age and public awareness of its awful potential consequences.

In the 1950s the systems analysts again found fertile fields for development and application in science-based industries. Governments also required tools to evaluate proposed expenditures and high technology investments, and cost-benefit analysis (CBA) became widely accepted, e.g. in defense. The aerospace industry combined these techniques into failure mode and effect analysis and reliability assessments. This was particularly important in that high reliability is essential in situations such as flight of spacecraft or aircraft where the consequences of failure are likely to be disastrous.

Today the same argument holds; the commercial implications of technical failures can be disastrous. The Flixborough disaster[2] meant the same to the chemical industry as the Tay Bridge disaster to civil engineering, in that it highlighted to the public the potential impact of chemical technology.

The nuclear industry was always associated in the public mind with atomic weapons, and the 1960s saw a growing but unsatisfied desire for effective public participation in planning, hence that period saw the rise of vigorous protest movements. The protests gained perhaps unwarranted support, in that government, seemingly both developer and regulatory authority, was open to criticism on the grounds of conflict of interest. The nuclear industry, confident in its capability and its future, held that its cost-benefit advantages and safety should be clearly demonstrable. Accordingly, criteria of acceptability were proposed based upon probability versus consequence arguments to demonstrate the care being taken, and were the basis of what came to be the techniques of risk assessment.

Since there was little experience with nuclear power plants on which to base empirical analysis (in contrast to practise in the insurance industry), hypothetical reliability models were developed to fill the gaps. A major problem arose when quantitative analyses based on hypothetical maximum credible accidents (MCA) predicted embarrassingly high costs and numbers of deaths (likelihood was not formally taken into account at that stage).

The worst postulated event approach can be recognised in the MCA philosophy currently employed by the chemical industry, which suffers from the same two problems:

(i) embarrassingly high consequences;
(ii) it reflects the biases of whoever decides what is credible.

In fairness, the approach has led to several useful pragmatic techniques, such as the Dow Index approach,[3] currently popular with Japanese industry.

In the USA the Environmental Protection Agency (EPA) in 1969 introduced a requirement for environmental impact statements (EIS), which were not initially intended to be a major feature of the legislation.

Environmental groups, however, were able to use them as a lever for protest, because they were critical points through which every planning decision had to pass. If the EIS was not credible or could be discredited, then the project was blocked. The same reasoning can be applied to the MCA approach to hazard impact assessment. The main environmental impact is from frequent low-level events or emissions. Hazard analysis entered the scene when the increasing scale of plants and financial investment gave rise to a series of events which were clearly very large and therefore unacceptable and highlighted the need to include in an impact assessment abnormal, infrequent, catastrophic events.

That led to a formal realisation, at least in some quarters, that the impact of a new development or technology was not just limited to MCA or an EIS but to an entire range of possible damaging impacts, some with less potential for damage but a greater probability and vice versa. That is, in fact, the current basis of hazard analysis.

FORMS OF HAZARD ANALYSIS

Process plant hazard analysis originated as a response to two main requirements:

   (i) the need to include abnormal events in environmental impact studies;
  (ii) the desire of designers to build safety into their plants in a rational way, rather than merely by routine application of established practice.

Environmental hazard and risk assessments are usually updated continuously throughout a project. For example, Imperial Chemical Industries (ICI) normally requires a different form of hazard analysis tailored to each distinct phase:

Phase I   conceptual
Phase II   outline plans
Phase III   detailed design
Phase IV   construction
Phase V   commissioning
Phase VI   audits (both in normal operation and after modifications)

We have adopted this as a basic model as it has proven effectiveness in a number of projects. The most appropriate form of hazard analysis to be employed in each case can also be dealt with in the above order.

Planning Studies

In this phase, impact on the environment is assessed and additional constraints in both design and location are identified. As only an outline design, or the design concept, is normally available at this stage (Phases I or II) the assessment must be based on realistic assumptions of detail. Where assumptions have to be pessimistic to allow for uncertainties in the final design, some of the results can be very constraining.

It has also become clear to us and to others producing these surveys that there is a real need to present results in a way that non-specialists can appreciate and understand and that is helpful to decision-makers (the 'so what?' test). This requires a considerable degree of condensation of the mass of results that can be generated from a risk analysis, and care must be taken to summarise them in a form that suits the problem in question (e.g. risk versus distance graphs can be used to derive the extent of safety exclusion zones).

Design Study

The full assessment is normally undertaken when the project design is, to all intents and purposes, finished and all flow-sheets, plant layouts, drawings, inventories, etc., are both complete and realistic. At this late stage in the project, when the detailed design data are available and can be frozen, there are at least two ways of assessing hazards formally. The first is through *check-lists*. There are a few examples of these in the literature and include techniques such as the Dow Index; they are also implicit in the application of relevant codes and standards, etc. The techniques essentially embody the sum of past experience in the industry and must be considered a necessary minimum step. For rigorous analysis of the design, however, the *hazard and operability study* (H&O)[4] is the best current method allowing a thorough review of possible process malfunctions.

The basis of the H&O technique is to examine systematically the purpose of each section, predict what could go wrong and ascertain how far the installation will respond in the way in which it was intended to perform; this can then be documented. This formal review is particularly important and necessary for new processes or developments. The H&O technique is well documented and much useful experience is available in the literature. Our own experience makes us enthusiastic but there is still scope for making the technique more cost effective, e.g. it may be necessary only to examine certain critical parts of the plant rigorously. People with established practical experience are essential to the H&O team, and the cost of dedicating an experienced team to an intensive review is considerable. An H&O study can, however, identify circumstances and events otherwise unforeseen even by experienced people.

As part of the questioning of the design, critical events and hardware, e.g. protective systems, can be identified, on which further detailed work may require *fault tree* techniques. These and other reliability and availability studies demand much time and effort but are well documented and proven in nuclear and aerospace industries. Their primary function is in evaluating reliability of complex systems.

H&O is the best systematic method for ensuring a thorough review of possible process malfunctions. Engineering design and proposals for construction procedures must also be checked at this stage, and if prior risk assessment has been made, the assumptions inevitably contained in it can be checked against actual design.

Construction

In the construction phase the hazard technique is essentially *quality control* and there is a need to provide formal procedures to ensure accuracy in materials specification, supply of testing programs, etc. Documentation must be foolproof and be seen to be both effective and necessary. It is basically obtaining *ground truth*, i.e. is what was intended actually built? Perhaps at this stage the risks involved in the construction process should also be examined, which is not done formally at present.

Commissioning

Again, procedures for commissioning and malfunctions to be expected can be derived directly from rigorous H&O. Ideally, the H&O team should review the study of equipment and procedures at start-up. The lessons from previous hazard studies can be advantageously incorporated into operating and emergency procedure manuals. In a very complex plant it may be useful to consider pretraining on a simulator developed from H&O data.

It is perhaps fair to comment that the hazards of commissioning are generally underestimated. For example, more people are killed by nitrogen (mistaken for air) than any other gas and the members of the highly competent and experienced commissioning team normally take important practical experience with them when they move on to the next commissioning, leaving the relatively inexperienced operations team aware mainly of routine.

Normal Operations

These require *process audits* which should be periodic, thorough and independent, as with financial audits, as familiarity inevitably breeds complacency. They should be structured either by a check-list or by reference to a list of possible failure cases. They can be used to check for any safe alterations or as a substitute for a risk assessment if none were done before the plant was built. But for a plant of any complexity, modifications or changes to the system should be accompanied by more rigorous H&O or risk quantification studies in their own right.

PREVIOUS HAZARD ANALYSES

There are few published examples of hazard analyses. Outside Scotland (see below) the obvious examples are the Canvey Island study[5] and the Rasmussen report.[6]

The Rasmussen report is a detailed quantitative analysis of nuclear reactor hazards and the risks involved, based largely on comprehensive and rigorous fault tree analysis of the reactor systems. Reduction of risk is the main purpose of the protective systems and a great deal of effort is spent on ensuring and demonstrating very high levels of reliability. A major problem with this type of study is the inadequacy of the basic failure rate data, due to the relatively short history of such plant in operation.

The Canvey Island study, on the other hand, was of an existing situation of public concern over the advisability of building additional high risk plants in the area. It was at the time the most rigorous definitive risk study ever undertaken of major hazards, explosions or toxic releases. Risk assessment was based largely on historical statistics of failures and little use of fault trees was possible or necessary. In view of the volume of analytical effort expended, it was unfortunate that the study did not include a thorough investigation of the implications of the mass of detailed results generated. In particular, the question of whether the calculated risk levels (which appeared to be quite high, but were almost certainly overestimated) should be considered satisfactory was only addressed in an oblique and ambiguous way.

REVIEW OF SCOTTISH PLANNING ASPECTS

Hazard analyses in Scotland are particularly useful examples because in many cases the initial conditions were roughly the same, so the historical way in which the approach to hazards evolved during the course of the North Sea saga can be seen.

In the earlier developments which were proposed, e.g. Nigg refinery, Sullom Voe terminal, attention was mainly focused on the environmental aspects of oil spills, etc., and hazard impact aspects were not formally recognised. After 1975 it was recognised that major hazards must be considered and that when siting installations separation distances may be critical. Until then this had been attempted using an MCA approach, which inevitably produced queries from industry on the levels of MCA design and much argument about the models and consequences. The consequences predicted from use of 'hypothetical models based on arbitrary data' gave high MCAs which were exploited by the objectors. In the Mossmorran development (Taylor, these proceedings) there emerged widely differing views as to the acceptability of the risk. This resulted in an impasse, where no clear guidelines were forthcoming, and worse still implied criticism that technical experts could not agree among themselves, nor could they convince lay observers that their predictions and calculations were valid.

The exercise highlighted the fact that decision-makers need

(i) planning tools for technological developments;
(ii) a logical or rational basis on which to make decisions;
(iii) guidance, particularly in local and regional planning (at the time authorities had a duty to identify developable land which required objective criteria for production of sensible plans).

In 1977 Cremer & Warner were briefed by the Highland Regional Council to produce *Guidelines for Layout and Safety Zones in Petrochemical Developments*. A logical procedure was therefore developed[7] which

(i) took a full conceptual description of the plant;
(ii) postulated a wide set of failure cases;
(iii) quantified consequences;
(iv) assigned frequencies;
(v) integrated the consequences and frequencies.

This procedure yielded objective criteria by which to establish distances between hazardous installations, and was subsequently developed into a computer program which produces contours of constant probability of exceeding a particular level of damage. Development of risk contour plots evolved in parallel with other risk studies such as by Science Applications Inc. at Covepoint in Maryland[8] and Batelle in their Gothenberg report[9] which both independently introduced an indication of risk at specific locations. The Cremer & Warner risk contour assessment technique has been employed successfully in Scotland in conceptual studies of a seabed-supported NGL plant at Pan Hope, Flotta, in an assessment of a proposed major LPG terminal in Shetland, and particularly in studies of the onshore developments proposed this year by the British Gas Corporation (BGC) to handle associated gas collected from North Sea fields.

The proposed St Fergus gas reception terminal, part of BGC's gas gathering project,[10] has been outlined by Dean (these proceedings). In the risk study commissioned by BGC, 630 failure events were postulated for the gas reception terminal and approximately 270 failure events for the downstream SNG plant. For each of these possible failure events, the consequences were computed taking account of the wind frequency pattern for the site (since this affects the direction of dispersion of vapor clouds), and detailed risk contour plots were then produced for various levels of damage. For example, the risk of serious explosion damage to process plant falls off rapidly with distances from the process modules on the terminal (Fig. 1) and so will not significantly affect the neighboring Shell plant to the south of the proposed SNG plant. The effect of adding in the risk from the failure events postulated for the SNG plant is illustrated in Fig. 2. Clearly the SNG plant has a much lower hazard impact. The analysis can be repeated for different levels of damage, e.g. risk contours for an explosion damage level sufficient to cause spillage from conventional oil storage tanks and serious (but repairable) damage to typical houses (Fig. 3).

Risk contour plots such as these are helpful in the planning procedures, and it is a relatively simple matter to consider the effects of variations in process concept or plant layout, and thus optimise the siting, design and layout of the hazardous development.

CONCLUSIONS

Modern hazard analysis is the result of an evolution of techniques which can be used for assessing and quantifying risk. Industrialists sometimes view it with a certain scepticism as a tax on new technology and as a

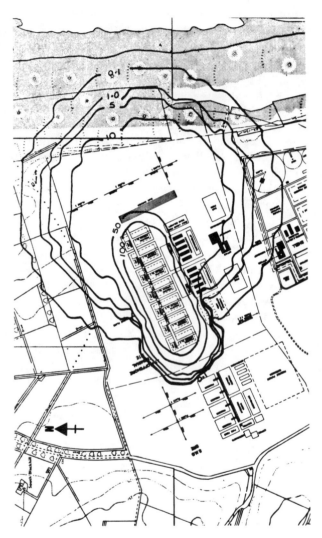

Fig. 1. Risk contour plot of predicted frequencies (occurrences/10^6 years) of serious explosion damage to process plant from incidents on the proposed St Fergus gas reception terminal. (Scale approx. 1:15 000.) Crown copyright reserved.

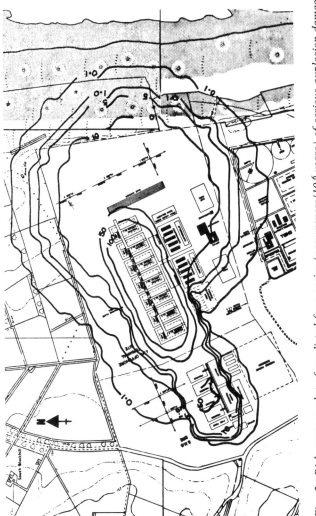

Fig. 2. Risk contour plot of predicted frequencies (occurrences/10⁶ years) of serious explosion damage to process plant from incidents on the gas terminal and SNG plant proposed for St Fergus. (Scale approx 1:15 000.) Crown copyright reserved.

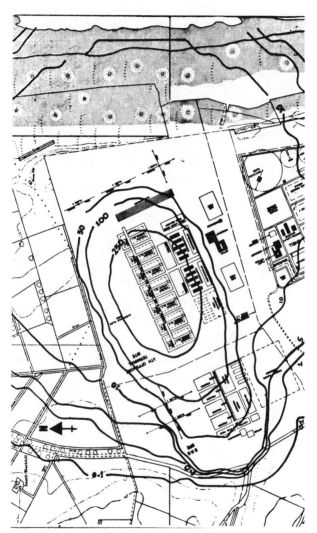

Fig. 3. Risk contour plot of predicted frequencies (occurrences/10⁶ years) of serious but repairable explosion damage to houses from incidents on the gas terminal and SNG plant proposed for St Fergus. (Scale approx. 1 : 15 000.) Crown copyright reserved.

further drain on available resources, financial and human. This attitude is understandable, because there is no immediate pay-off, but it should be remembered that hazard analyses often lead to more reliable operation in service, and can help avoid expensive planning delays.

Hazard analysis does seem to be fulfilling a real need of local and central government. New legislation in the UK and EEC will make further demands on government decision-makers and industry in this area.

It is crucial that hazard analysis should be seen as a tool not only for professionals in industry and government but as an aid for communicating technical safety issues to the public. It can and should be used in public debate to provide a logical structure within which the arguments of all sides can be evaluated and compared.

The techniques of hazard analysis are still evolving. This is an encouraging sign particularly as we are not concerned solely with short-term economic gains but long-term goals, which are the essential basis of sound planning.

REFERENCES

1. Nash, J. R. (1979). *Darkest hours.* Wallaby Books, New York, USA.
2. *The Flixborough disaster.* Report of the Court of Inquiry, Chairman: R. J. Parker, HMSO, UK (1975).
3. American Institution of Chemical Engineers (1973). *Hazard classification and protection—fire and explosion safety and loss prevention guide.* CEP Technical Manual.
4. Anon. (1977). *A guide to hazard and operability studies.* Chemical Industries Association.
5. Anon. (1978). *Canvey: an investigation of potential hazards from operations in the Canvey Island/Thurrock area.* Health and Safety Executive, HMSO, UK.
6. US Nuclear Regulatory Commission (1975). *Reactor safety study: an assessment of accident risks in US commercial nuclear power plants.*
7. Cremer & Warner (1978). *Guidelines for layout and safety zones in petrochemical developments.* Prepared for the Highland Regional Council. Ref. C. 2056.
8. Anon. (1978). *Risk assessment study for the Cove Point, Maryland LNG facility.* Science Applications Inc.
9. Anon. (1978). *Risk assessment study for the Harbor of Gothenberg.* Batelle Institute, Frankfurt, West Germany.
10. British Gas Corporation and Mobil North Sea Ltd (1980). *A North Sea gas gathering system.* Energy Paper No. 44. Department of Energy, HMSO, London, UK.

22

A Company's Experience of Environmental Protection Requirements at Continental European Coastal Sites

G. NORCROSS

ICI Europa Ltd, Everberg, Belgium

ABSTRACT

Since 1962 ICI has established large-scale manufacturing operations on the coasts of Holland, France and Germany. Until recently some smoke and water pollution in the vicinity of large-scale industry was accepted as normal and public concern about industrial safety was confined to occasional newsworthy accidents. Establishment of the continental ICI sites, however, has coincided with growing public awareness that responsible environmental management is needed to prevent irreversible damage to the balance of nature and most countries are now developing legislation to protect the environment.

Approaches to environmental protection reflect local conditions and national characteristics and traditions. Initially the objective of environmental protection was preservation of ecosystems from destruction, but later the main concern was the safety and comfort of human beings. The chemical industry is increasingly subject to legislation designed to protect the environment and the public. It is a relatively safe industry and ICI has adopted measures to ensure that it operates in a safe and environmentally acceptable manner.

DEVELOPMENT OF ENVIRONMENTAL LEGISLATION

Imperial Chemical Industries (ICI) has been building chemical plants on continental Europe for 18 years. In 1962 the word 'environment' had not yet been applied in connection with effluent treatment. It was still accepted by the public at large that a certain amount of pollution of air and water was to be expected in the vicinity of industrial activity. Governments, however, were already controlling some forms of pollution by law in certain areas. For example, the burning of coal was prohibited in London in the 13th century (1273) because it was considered that the smoke was a danger to health. The Alkali Act, upon which the control of pollution from the chemical industry in the UK is still based, was passed in 1863 and consolidated into its present form in 1906, and the Clean Air Act, which has had such a beneficial effect in removing smoke, fog and grime from big cities in Britain, dates from 1956. Industry, too, being a part of the society in which it operates, for the most part took a positive attitude and was already minimising pollution in many ways, often without waiting for a specific law or a demand on the part of the authorities. As an example, one may cite the problem of foam on rivers in the 1960s. The foam sometimes built up at weirs to inconvenient and even risky proportions and was blown around by the wind. When it had been established, with the help of industry, that the foam was a result of the rapidly increasing use of non-biodegradable synthetic detergents, the detergent industry and the chemical industry throughout Europe and North America developed biodegradable detergents before legislation was passed banning the use of the non-biodegradable variety.

Thus, well before 1962, although the news media and the public were not greatly concerned about pollution abatement, industry was controlling pollution in accordance with the laws, with the existing state of knowledge and with the social climate.

The 18 years of ICI's production activities in continental Europe have coincided with growing awareness that the world's resources are not boundless and that the ability of nature to assimilate pollution is also limited. These notions were popularised by intellectuals in organisations such as the Club of Rome and by books such as *Silent Spring*.[1] They were perhaps brought home to the public in a graphic way by the pictures taken by the astronauts of planet earth floating in space. Evidence of the risk to a fragile planet came from the announcement by Thor Heyerdahl that he had met oil pollution in the middle of the oceans and also from the rapidly increasing sensitivity of analytical tools which enabled traces of man-made chemicals to be detected thousands of miles from their sources. Public awareness of environmental matters was also enhanced by the attention paid by the news media to the local ecological consequences of major oil spills, particularly around the coasts of Europe, e.g. from the *Torrey Canyon* and the *Amoco Cadiz*.

In the years following World War II, the population of Europe was increasing and individuals were enjoying a rapidly improving standard of living. This was achieved by factories becoming larger and more numerous. Each worker had more energy under his control, was producing more and therefore using more raw material, handling bigger inventories of intermediate products and producing more byproduct, i.e. more potential pollution. Pollution was no longer localised but was affecting wider areas and crossing international boundaries. Well known examples are the 'acid rain' in Sweden, ascribed by the Swedes to sulfur dioxide from chimneys in the UK and Germany, and the pollution of the Rhine by countries upstream, which causes problems in Holland where Rhine water is vital both as a source of drinking water and to stop the infiltration of salt water from the sea.

Nature's ability to cope with pollution would be overwhelmed and ecosystems would be destroyed over wide areas if urgent steps were not taken, as Captain Cousteau warned about the Mediterranean Sea. These developments coincided with public awareness of the finite nature of the earth and growing demands at the same time for the preservation and restriction of amenities such as absence of smog, rivers with fish in them and water fit for bathing.

It is not surprising, therefore, that since 1962 there has been progressive tightening of environmental restrictions in the UK and a great deal of environmental legislation throughout Europe, both within countries and by the European Community, to control the activities of producing industry, starting with the limitation of discharges to surface water and later extending to ground water, gaseous discharges, noise, solid waste disposal and environmental impact in the widest sense.

Similarly, in 1962 neither the public nor the authorities were nervous about the safety of the chemical industry. As late as 1973, the EEC Commission had no plans to legislate about safety in the chemical industry because it was considered to be a safe industry; and indeed it was. The average worker in the chemical industry is safer than the average construction worker, sea fisherman or coal miner, and the chemical industry is becoming safer still.[2-6] However, the shock of such isolated incidents as those at Flixborough (1974) and Seveso (1976) has transformed the public image of the chemical industry into that of a hazardous industry. Here again, the political authorities have been developing legislation at both national and European Community level, designed to ensure as far as possible the safety of the public from the risk of explosions and toxic releases.

The approach to environmental protection varies from country to country, naturally reflecting the different local conditions such as the density of industrialisation, the proximity of industry to the sea, the length and rate of flow of rivers and the direction and strength of prevailing winds. The approach also varies, however, in a way that reflects national characteristics and traditions and the European Communities encounter these differences in developing EEC directives in the environmental and safety fields. One example is the belief in some countries, notably in the UK, that the most cost-effective means of protecting the environment is for the authorities to agree environmental quality objectives with effluent dischargers. In other countries, the alternative uniform discharge standards are preferred with just as much strength of feeling.

The approach to environmental protection is also changing. At first the main thrust was to diminish pollution, particularly water pollution, to protect ecological systems which were threatened with destruction if nothing were done. Later, the main concern of the public was the safety of people, who must be protected from toxic chemicals in the atmosphere, in drinking water and in fish used for food. As rivers became cleaner and the air over cities improved, the public took it for granted that industry would protect water and air from gross pollution. The objective of environmental control is now being broadened to protect the comfort and amenity of the public, and the public itself is being brought into the administrative process of deciding environmental controls and licensing the construction of new factories and facilities.

ENVIRONMENTAL CONSIDERATIONS FOR ICI'S EUROPEAN ACTIVITIES

In 1960 ICI had most of its production facilities in the UK and substantial manufacturing activity all around the world in the countries of the British Commonwealth. On the Continent of Europe, the Company had no large manufacturing units but an interest in a few small sites concerned, for example, with warehousing and with the formulation and packaging of pesticides. In that year the ICI Board in London decided that even though the UK was not in the EEC at that time, nevertheless the six countries making up the EEC would be important markets for ICI products in the future and that production plants should be established there.

The Netherlands

As the first step the ICI Board decided to establish an integrated chemical complex at Rozenburg in the Europoort area of Rotterdam in The Netherlands. Rozenburg had excellent transport connections for feedstock supply and dispatch of finished products and was surrounded by a type of industry within which the ICI plants could be integrated, and where chemical production activities could be supported. The area houses five major refineries, with a combined annual capacity of 85×10^6 tonnes and has the largest concentration of chemical plants in Holland.

At the time of the ICI decision to build there, much of the land on which the Rozenburg works now stands was below sea level. The development of the 147 ha site was carried out by the municipal port authorities of Rotterdam, who had to raise the ground level by 4–6 m. This required the pumping of some 14×10^6 tonnes of sand, and the 'hole' that was thus created became the harbor. The first pile, for what has become ICI's biggest European complex so far, was driven in May 1962 and the first plant to be erected came on stream in 1963.

Environmental matters were not a major problem. The first plants were for production of plastics and polymers from chemicals mostly imported by ship from ICI plants in the UK and, with the up-to-date technology, very little effluent was created. The Rotterdam Port Authorities were satisfied with the arrangements ICI proposed for effluent disposal, namely to treat domestic effluent in septic tanks and remove oil from the plant effluents. The site effluent was then discharged into the harbor which is connected to the sea via a 6 km shipping channel. By the end of the decade, eight plants were in operation; the effluent discharge to the harbor was inoffensive but the organic pollution in it had reached the level of that expected from a town of 20 000 inhabitants.

In 1970 plans were made to manufacture chemicals at Rozenburg for the first time, the new plants alone producing organic pollution equivalent to that from 40 000 people. It was agreed with the authorities that effluent treatment plant would be installed on the site to reduce the expected total of 60 000 population equivalents (p.e.) to the existing level of 20 000 p.e. and the new chemical plants and the effluent treatment plant were brought into operation at the same time.

The Rozenburg management was helped in its negotiations with the authorities, and in the design and justification of the effluent treatment plant by chemical process specialists in what is now the ICI Organics

Division in Manchester, by the Engineering Department at the ICI Europa Headquarters in Brussels and by the ICI Brixham Laboratory (Devon). The 70 scientists and technicians at Brixham provide a service for the whole of ICI in marine surveying, the effects of pollutants on fish and other aquatic life and the design of water effluent treatment plants. The Laboratory also carries out consultancy work for companies and authorities outside ICI and enjoys a worldwide reputation for its technical competence.

The main item in the effluent treatment plant installed at Rozenburg was a 'Flocor' trickling filter. The 'Flocor' system had been developed at the Brixham Laboratory and was related to the system traditionally used in sewage works, where the sewage is distributed by rotating perforated pipes over broken stone in circular beds. The bacteria which destroy the pollution grow on the broken stone, or on the plastic packing in the case of the 'Flocor' tower. The 'Flocor' system had already been applied successfully in many plants treating food factory waste and town sewage but this was one of the earliest applications in chemical plant and was, in effect, a full-scale experiment. Being a living system, the bacteria could be disturbed by toxic chemicals accidentally discharged by occasional process irregularities during the running-in period of the new production plants; when the bacteria were thus disturbed, very little purification of effluent took place and a highly polluted effluent reached the harbor until the bacteria re-established themselves.

The Brixham Laboratory scientists agreed with the management of the Rozenburg factory that if the treatment plant were to achieve its objective fully, it must be extended by further biological treatment and by removal of the sludge generated in the process. At the same time, the authorities offered a grant towards the cost of the extensions because they liked the design which ICI proposed and because the extended effluent treatment plant would be big enough to bring the effluent load to the harbor well below the original target figure of 20 000 p.e. The extended effluent treatment plant was started up in 1977 and has been operating satisfactorily since then. The grant offered by the authorities was obtained by them from the effluent taxes which are levied in Holland on industries and municipalities which discharge polluted effluent to natural waters. These taxes are calculated on the amount of pollution discharged but nevertheless are intended as a source of revenue to cover the cost of pollution control administration and to provide subsidies for effluent treatment plants such as that at Rozenburg. The taxes are not intended to be an incentive to polluters to install effluent treatment plant; the reduction in tax achieved by installing effluent treatment plant is small compared with the cost of operating the treatment plant.

Although the most expensive and most visible single environmental protection item on the site, the water effluent treatment plant is but one of many measures taken by the ICI works to protect the environment. There is continual contact with the authorities, who expect ICI to use the most appropriate environmental protection measures at all times, in a way similar to the 'best practicable means' policy familiar in the UK. There is mutual respect for technical competence but no sign of the too-easy relationship suspected by the action groups to exist between industry and the authorities. The authorities are firm when necessary, as when they shut down[7] a petrochemical plant in the Europoort area for several weeks in 1970 because of the nuisance and anxiety caused to the public by high level flaring of waste hydrocarbon gases during plant upsets; and as when they fined[8] a chemical company for failing to report promptly an emergency discharge of gas and smoke in 1974. The local inspectors are supported in negotiations with industry where appropriate by technical experts at provincial and central government level and also by the well known government supported TNO (Applied Scientific Research) Laboratory at Delft.

Second in size in the Rotterdam-Europoort area only to Shell, ICI plays a leading role in the Rotterdam chemicals industry. ICI's work in the UK on the development of systematic analysis of safety and risks[9-11] and of hazard and operability study techniques[12, 13] (see also Slater *et al.*, these proceedings) has given the Company a very high reputation[14] and the authorities regard the work as a good example for other industries. Through Rozenburg, the Company has made a big contribution to the development of the unique Europoort/Botlek Foundation, an organisation of 65 firms in the Rotterdam-Europoort area, which has been set up to deal on a collective basis (and more efficiently than could be done individually) with problems that might face any of its members. Probably the most important activity of the group is the extensive service it has established for environmental control and accident prevention, and for mutual support in case of fire or other calamity. The activities, the equipment and manpower resources of the member companies have been pooled into one coordinated unit, with the most modern radio communication system closely integrated with the police and public fire brigades. The main control room of this system is located at the ICI Rozenburg works. Another important facility is the common medical center located near to the ICI site.

France

In 1970 ICI was seeking a second continental site, this time in southern Europe, and was offered land in the Fos industrial zone west of Marseille. Unlike Rotterdam, there was very little industry already in operation in the industrial zone but there were several established oil refineries 25–50 km away. The Fos industrial zone itself was scheduled for vigorous development, with support from the central government, and there were plans for chemical factories, steel and non-ferrous metals production. Roads and port facilities were being developed and water supplies, electricity and telecommunications assured.

The Fos industrial zone was seen by the government as providing employment in the region, to replace jobs lost with the decline in agricultural employment and the decline of the port of Marseille following loss of the north African colonies and closure of the Suez Canal in 1967. Local people were rather sceptical, since the heavy industry planned for the industrial zone seemed to offer jobs to immigrants from other parts of France rather than to locals, and the downstream light industry would be established 100–300 km away. Meanwhile the industrial zone would transform the area in an unwelcome way; environmentalists also feared the effect of heavy industry on the nearby Camargue conservation areas. Because of these fears, President Pompidou insisted that the Fos zone should be a model of environmental cleanliness. The concept of the 'environment' was by then in vogue and the French government had announced in June 1970 the '100 measures for the protection of the environment'.

The first plant to be started up by ICI at Fos in 1972 took ethylene by a high pressure pipeline from a cracker 25 km away and polymerised it to produce polyethylene. This is a relatively clean operation and it was not difficult to meet the environmental requirements. Domestic waste water was treated in septic tanks and rain water was freed from oil contamination before discharge to the drainage canal leading to the harbor. Air purity was a serious concern; the ICI works was asked to hold stocks of fuel oil containing less than 1·5% sulfur so that the boilers could stop burning the normal fuel oil if the works were warned that the sulfur dioxide concentration in the district was rising too high. Small amounts of ethylene gas are continuously purged from a polyethylene plant and, being innocuous when diluted by air, are often blown off to the atmosphere. At Fos, however, the authorities feared that under certain weather conditions ethylene could contribute to the formation of petrochemical smog and ICI agreed to install equipment to collect the purged ethylene and burn it under the steam boiler.

In France, construction permits and operating permits are issued by the Prefect of the Department, who is appointed by the central government. The Prefect takes advice from local public service departments (health, fire, police, agriculture, transport) and, particularly in a case like ICI Fos, from the Mines Inspectorate (Service des Mines) which is somewhat akin to the UK Factory Inspectorate and Alkali Inspectorate combined. In practice, a company like ICI must first convince the Mines Inspectorate of the appropriateness of its environmental and safety measures and then the Inspectorate helps to prepare a document for exposure to the public on town hall notice boards. In the Fos and Etang de Berre area, the Prefect set up a body in 1972 called Secretariat Permanent pour les Problèmes de Pollution Industrielle (SPPPI) to advise him about environmental matters. It is chaired by the local head of the Mines Inspectorate and contains people from the public administration, industry and also elected people such as mayors. Thirteen industrial firms formed an association, l'Association des Industriels de la Région de Fos–Etang de Berre pour l'Etude de la Prevention de la Pollution (AIRFOBEP), in the same area, to enable them to talk to the SPPPI with one voice and to cooperate in environmental and safety measures. As at Rozenburg, ICI is taking an active part in AIRFOBEP. The Mines Inspectorate maintains continual contact with the ICI works at Fos, paying particular attention to water effluent, sulfur dioxide generation and disposal of solid waste and, as at Rozenburg, expects ICI to use the most up-to-date environmental protection measures.

West Germany

By the time ICI came to seek another major manufacturing site on the Continent in 1975, environmental considerations and questions of public safety had become major concerns, not only to industry but also to the authorities and the public.

Out of an original list of 25 sites, a site was chosen in early 1977 at Wilhelmshaven on the North Sea coast of West Germany. The available area was at the northern end of a long, narrow (2 km wide) strip of land reclaimed by diking in mudflats, as at Rozenburg, and raising the ground level by pumping in silt. Wilhelmshaven is a naval port, created from nothing in the 19th century on the narrow Jade Bay. There is very little industry in the area. With the decline in dockyard employment, the

Lower Saxony Land Government helped the town over the period 1965–75 to reclaim the strip of mudflats to the north of the town for industrial development. By the time ICI took an interest, there were in operation only a power station and a small chemical works at the southern end, near the town, and the Mobil Oil refinery towards the northern end of the strip. Significant points to note are

(i) the Jade Bay is shallow and has only a slow exchange of water with the open sea;
(ii) the extensive remaining mudflats are a nature reserve;
(iii) there is mussel fishing for food in the bay;
(iv) the northern boundary of the ICI site would coincide with the administrative boundary of Wilhelmshaven; the district of Wangerland to the north has no industry and lives from farming, fishing and catering for holidaymakers.

In Holland and France, the law says in effect that a factory must use the most up-to-date environmental protection measures technically available and economically supportable. In Germany, however, there is the difference that the Federal Environment Law of 1974 was followed by Technical Guidelines for Noise Control and for Clean Air (known as T A Larm and T A Luft) which contain limiting values expressed in numbers and which have the force of law. ICI therefore had to forecast what gaseous discharges and what noise levels it would be able to achieve, not only from the first three plants and associated services but also from the other 20 or so unknown plants which the company would expect to build during the ensuing 15–20 years, to ensure that we could meet these numbers.

The question of noise gave the greatest headache because there were housing areas within about 1 km of our nearest plants; in Germany one is not allowed to impose more than 35dB on an area designated as a purely residential zone and that is very quiet indeed. The forecasting of noise imposed at a distance is not an exact science and the formulae and assumptions used by the ICI noise engineers were different from those used in Germany. Detailed comparisons of the mathematics and basic assumptions of the two systems were made by ICI noise engineers and the German noise consultants and comparative measurements made at a factory in England in order to establish the difference in decibels between the forecasts obtained by the two systems. ICI management could then be satisfied that it would be possible to satisfy all foreseeable environmental requirements at Wilhelmshaven.

Two features in particular of the German permit system distinguish it from other systems. One is the liberal use made of Gutachter (expert consultants) and the other is the amount of information the German authorities give to the public. Both features appear to stem from a wish to demonstrate that there is no friendly arrangement between the authorities and the company to the detriment of the public. At Wilhelmshaven, ICI's description of the first plants ran to nine large volumes. The technical resources of three ICI divisions and the central laboratories and services were marshalled to provide, for plants which had not yet been designed, details of all conceivable environmental effects of the plants, that is, effects on people and nature outside the factory fence. These were compiled, translated into German and printed in a matter of a few weeks. Copies could be consulted by the public in the town hall for 2 months and individuals were able to ask for photocopies of pages which interested them. Written objections were dealt with at public meetings, but by that stage ICI had convinced the authorities that its proposals were reasonable and it was the authorities who in effect defended the project at the public meetings.

One of the nine volumes for the information of the public contained five detailed reports by independent experts on such aspects as the expected concentrations of gaseous effluents from the factory at specific villages in the area, calculated by a computer model, the safety of the public who might breathe traces of such gaseous effluents and the safety of the public should fire or explosions occur. Although the choice of experts was approved by the authorities, their fees were paid by ICI and, to ensure impartiality, the authorities appointed other experts to comment on some of their opinions. In reassuring the authorities and the public about the innocuous nature of the traces of gaseous effluents which might reach villages in the area of the factory, we were fortunate in having behind us the ICI Central Toxicology Laboratory (CTL) at Alderley Park near Manchester where over 300 scientists and technicians are employed. CTL is one of the largest laboratories of its kind in the world and is responsible for advising all parts of ICI about the effects of chemicals on people and methods of safe handling. Apart from testing the effect of chemicals on animals such as rats, mice and guinea pigs, the laboratory is gaining an international reputation for its *in vitro* toxicity testing, using bacteria and cell cultures, and is collaborating with the US Environmental Protection Agency on a program concerned with short-term tests for carcinogenicity.

The first ICI plants at Wilhelmshaven are about to start production.

These world-scale plants are an awe inspiring sight but, by skilful landscaping, they have been made unobtrusive and largely invisible from nearby villages. ICI is confident that it will shortly be an accepted part of the local scene, even to holidaymakers and those who cater for them.

CONCLUSIONS

ICI's ability to find sites and to build plants which comply with the environmental requirements in different continental European countries, always to a tight timetable, depends not only on technical expertise available from the company in the UK but also on the small international team of ICI engineers and technologists on the Continent, mostly in the Brussels headquarters of ICI Europa, who facilitate contact with technical organisations in different countries and avoid misunderstandings. The fluency of many of them in two, or more, foreign languages is obviously valuable in itself but, more important and perhaps less obvious, it leads to that vital ability to cope with the complete change in outlook and way of doing things which occurs as one crosses any national frontier.

REFERENCES

1. Carson, R. (1962). *Silent spring*. Hamish Hamilton, London, UK.
2. Kletz, T. A. (1977). The risk equations—what risk should we run? *New Scientist*, May 1977, 320.
3. Dutch Central Bureau of Statistics (1975). Statistiek der Bedrijfsongevallen. *Statistisch Jaarboek.*
4. Gibson, S. B. (1976). The use of quantitative risk criteria in hazard analysis. *Journal of Occupational Accidents*, 1-85-94.
5. Waterhouse, P. (1980). In *Measurement of safety performance*. National Health and Safety Conference 1980. Reprints available from Victor Green Publications Ltd, 106 Hampstead Road, London NW1 2LS, UK.
6. Anon. (1979). *Sicherheit in der chemie*—BASF Symposium, 15 November 1978. Wissenschaft und Politik, Berend von Nottbeek, Cologne, West Germany, pp. 19, 64, 268.
7. Anon. (1972). *European Chemical News*, 10 March, **21** (523), 6.
8. Anon. (1975). *European Chemical News*, 14 Feb., **27** (674), 12.
9. Lawley, H. G. (1976). Size up plant hazards this way. *Hydrocarbon Processing*, **55** (4), 247.
10. Kletz, T. A. (1978). Practical application of hazard analysis. *Chemical Engineering Progress*, **74** (10), 47.

11. Lawley, H. G. (1980). Safety technology in the chemical industry: a problem in hazard analysis with solution. *Reliability Engineering*, **1** (2), 89.
12. Lawley, H. G. (1974). Operability studies and hazard analysis. *Chemical Engineering Progress*, **70** (4), 45.
13. Kletz, T. A. (1977). Evaluate risk in plant design. *Hydrocarbon Processing*, **56** (5), 297.
14. Knowlton, R. E. & Shipley, D. K. (1977). *A guide to hazard and operability studies*. Chemical Industry Safety and Health Council, Chemical Industries Association, 93 Albert Embankment, London SE1 7TU, UK. (Published in German as *Der Storfall im chemischen Betrieb* by the Berufsgenossen-schaft der chemischen Industrie, Postfach 10 14 80, D-6900 Heidelberg, West Germany.)

23

Protection of the Marine Environment of Scapa Flow and Sullom Voe

C. S. JOHNSTON

Institute of Offshore Engineering, Heriot-Watt University, Edinburgh, UK

ABSTRACT

Oil-related developments at Scapa Flow, Orkney, and Sullom Voe, Shetland, involve landfall of major pipelines from the North Sea to oil-loading marine terminals on industrially undeveloped islands. Occidental operates the Flotta (Orkney) terminal and British Petroleum the larger Sullom Voe terminal.

To minimise impact on the relatively unpolluted marine environment, environmental policy was incorporated in all aspects of development from planning through construction to operation. The uncertainty of potential effects required that environmental programs be developed cooperatively with the operators and statutory bodies, i.e. application of the 'best possible means' philosophy integral to much UK pollution control legislation.

Environmental risk from such oil-related activities and organisation of relevant programs is considered, with emphasis on impact assessment, pre-operational studies and subsequent monitoring strategy. The roles of operators, local authorities, central government, environmental consultants and others are discussed.

INTRODUCTION

There are four main and interrelated areas at risk in the marine environment: amenity, particularly beaches; resources, e.g. fisheries; bird life;

marine ecosystem(s). The direct effects of oil spillage on beaches and fisheries (e.g. net fouling) clearly represent a major definable economic impact of oil, and are well reviewed elsewhere. Similarly, seabirds, although part of a marine ecosystem, usually attract direct and not surprising emotional attention from scientists, press and thence the general public. Seabirds usually represent the only group of marine organisms under major direct threat from both acute and chronic spillage of oil.

The most complex impact to assess is that of oil on a general marine ecosystem. It is, therefore, also difficult to monitor, and is the focus of the present paper.

The term oil is used to describe crude oils (which vary considerably), petroleum products (e.g. diesel), single or known mixtures of hydrocarbons and sometimes non-petrolic material. Thus, unlike most other pollutants (e.g. metals, pesticides) oil is a complex mixture, varying and frequently unresolved. A crude oil contains hydrocarbons and non-hydrocarbons with the former usually representing over 75% of the total mixture. Hydrocarbons in turn can be divided into four main categories;

    (i) normal alkanes (paraffins);
   (ii) branched alkanes;
  (iii) cycloalkanes (naphthenes);
  (iv) aromatics.

The various proportion of each of these fractions, and the variation in balance of molecular weights of compounds within each fraction, greatly alters the behaviour, fate and toxicity of a spilled or deliberately discharged oil. To further complicate the problem the nature of, for instance, a spilled oil changes immediately it enters the environment, from the combined action of evaporation, dissolution, decomposition, etc., i.e. what is generally termed weathering.

In assessing the risk to a marine ecosystem it is impossible to do any more than make some key, perhaps sweeping, generalisations as already quoted:[1]

    (i) the toxicity of oil (extracts) appears to be a function of soluble aromatic hydrocarbons, particularly the alkyl derivatives of benzene, naphthalene and certain triaromatics;
   (ii) lethal toxicity ranges from 0·1 to 100 mg litre^{-1} (ppm) for adult organisms;
  (iii) larvae are 10–100 times more sensitive than adults;
  (iv) sub-lethal effects have been clearly demonstrated with very low

(ppb) levels of aromatic hydrocarbons;

(v) several polynuclear aromatic hydrocarbons have strong muta-
genic/carcinogenic activity;

(vi) certain metabolic products may be more toxic than their hydrocarbon precursors.

Against these very general comments regarding oil in the marine
environment, environmental policy (and studies) at two major new
onshore developments in the UK is considered in this paper.

Several of the North Sea oil developments involve shipment of oil
from offshore production facilities by submarine pipeline to onshore oil
storage, processing and tanker loading terminals. Frequent reference has
been made elsewhere in these proceedings to two such major terminal
facilities: Sullom Voe (Shetland) and Flotta (Scapa Flow, Orkney). In
considering the impact of such developments on the environment, most
of the papers have concentrated on environment as perceived by the
planner, i.e. a land-based view with emphasis on such considerations as
visual and social impact. This paper concentrates on the marine environment, with particular attention on the potential impact from routine operations rather than the possible threat from accidents such as major oil
spillages, which is extensively reviewed elsewhere.

Throughout design, development and operation of the Flotta terminal
the Occidental Group has retained the Institute of Offshore Engineering
(IOE) as marine environmental consultants and more recently British
Petroleum as operator of the Sullom Voe terminal has retained IOE as
contractor to undertake statutory marine environmental monitoring
studies. This paper concentrates on experience relevant to the Flotta
terminal with comparison where appropriate to Sullom Voe, with general
reference to problems of marine environmental monitoring. Environmental aspects of both terminal operations have been extensively reported
elsewhere[1-5], but it is important to identify certain key points to ensure a
balanced consideration of the total environment.

THE FLOTTA TERMINAL

The Occidental Group discovered and successfully brought on stream
the Piper and Claymore fields in the North Sea. Oil produced in these
fields is piped to a terminal on the island of Flotta which lies to the South
of Scapa Flow, Orkney (Fig. 1). This terminal provides oil storage, gas
separation/liquefaction and subsequent tanker loading facilities.[6] The

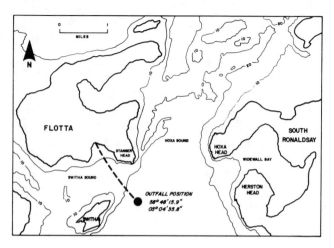

Fig. 1. Flotta oil terminal discharge site in Hoxa/Switha Sounds. (From ref. 5.)

operation of such a terminal presents the threat of oil pollution of the marine environment from four main sources:

(i) rupture of pipeline(s) in terminal environs;
(ii) major tanker spillage, particularly in approaches to the terminal;
(iii) oil spillage at loading facility (e.g. single-point mooring towers in Scapa Flow), particularly minor incidents;
(iv) operational discharges, via effluent discharge line.[3]

From initial planning and throughout construction and operation of the terminal there has been direct strong but positive communication between the Orkney Islands Council (formerly Orkney County Council) and the operator, the Occidental Group, with the Orkney Islands Council exercising maximum control over protection of the marine environment within the terms of the Orkney County Council Act 1974 and in its role as official pollution control authority. Direct statutory control over discharges and potential damage to the marine environment is the responsibility of the Secretary of State for Scotland under exemption terms of the Prevention of Oil Pollution Act 1971, the Scottish Development Department (SDD) receiving the scientific support of the Department of Agriculture and Fisheries for Scotland (DAFS). Responsibilities and restraints set upon the operator have been clearly identified from initial consent for terminal development and have left scope for both

formal and informal liaison and discussion between interested environmental bodies, notably

(i) Orkney Islands Council with their marine environmental consul-
    tants, the University of Dundee and their general environmental
    consultants Moira and Moira;
(ii) Occidental Group with their marine environmental consultants,
    the Institute of Offshore Engineering, and their general environ-
    mental consultants, W. J. Cairns and Partners;
(iii) Nature Conservancy Council (NCC);
(iv) Royal Society for the Protection of Birds (RSPB);
(v) DAFS;
(vi) Seals Research Unit of the Natural Environment Research
    Council.

The key component of the Occidental Group environmental protection
policy was prevention rather than restorative measures, with considerable
attention being given to effluent treatment and optimum discharge conditions.

Impact Assessment

Although there is no statutory requirement for an environmental impact
assessment (EIA) for such a UK development, both the Orkney Islands
Council and the Occidental Group commissioned major impact studies
to identify risk to the environment and form a basis for decisions regarding terminal construction and operation within an overall government
strategy set to ensure a major supply of oil for the nation from North Sea
developments.[7-9]
    A key feature emerging from these studies was the need to ensure

(i) thorough understanding of nature and levels of (oil) contaminated
    water required to be treated/discharged from terminal operations;
(ii) installation and operation of effluent treatment plant using best
    available technology to minimise need to discharge contaminated
    water;
(iii) selection, installation and maintenance of best practical effluent
    monitoring instrumentation/methods;
(iv) design and siting of effluent discharge line to give maximum prac-
    tical dispersion of discharged effluent;
(v) design and undertaking of environmental pre-operational and

monitoring program to check for no/minimal damage to receiving
environment and provide a feedback for improved effluent
treatment.

These points have been thoroughly considered elsewhere against a
generally improving understanding of oily water discharges.[3,4]

During normal operations there is a requirement to discharge contaminated water to the sea from the effluent treatment facilities at the terminal. The effluent is derived from oily water from

  (i) ballast from tankers;
 (ii) desalter/production water;
(iii) operational drainage from the terminal.

Present accepted oil:water separation systems depend primarily on a
physical separation of oil from water. However, studies in this laboratory
and elsewhere suggest that some effluent, particularly from desalter/pro-
duction water, can be enriched with lower molecular weight hydrocarbons, notably aromatics. Currently accepted monitoring routines (e.g.
infra-red) can grossly underestimate aromatic enrichment. Since such
aromatics appear to be the most toxic components of oil, greater attention should be given to ensuring a good location for final effluent discharge
than to attempt to reduce apparent oil concentrations in an effluent, e.g.
such statutory limits as 25 ppm total oil, when such measures bear little
or no relation to the likely toxicity to marine life.[1]

Initial studies also suggested the possibile presence of pollutants other
than oil, including material from ballast if picked up by a tanker at a
polluted port (e.g. Rhine), heavy metals in production water, and residual
chemicals employed in offshore/terminal operations (e.g. biocides).

Following detailed discussions between the Occidental Group, Orkney
Islands Council and the SDD it was agreed that treated effluent should be
discharged in the Hoxa Sound, south of Stanger Head, into a region of
considerable tidal flow. The final site, selected by Orkney Islands Council,
was due east of the island of Switha (Fig. 1).

Pre-operational Studies

An extensive range of shore studies was initiated by the University of
Dundee in 1974 (commissioned by Orkney Islands Council) involving a
quantitative transect survey and selected sites for population dynamics
studies.[10] In addition plankton and limited benthic studies were undertaken in Scapa Flow.

The RSPB carried out extensive impact and pre-operational studies of key bird species around Scapa Flow, which extended to feeding studies with financial assistance from the NCC. The Seal Research Unit carried out a number of population studies of grey and common seals in the area of Scapa Flow. As consultants to the Occidental Group, IOE undertook major studies of surface macrofauna and kelp infauna in the immediate turbulent environs of the effluent discharge but concentrated major effort on a chemical baseline program. Statutory control on effluent discharge required DAFS to undertake hydrographic, benthic and chemical studies within the discharge site.

Monitoring

Much of the general biological survey initiated prior to operation of the terminal has continued, including that by the University of Dundee on behalf of Orkney Islands Council.

Within the exemption conditions of the Prevention of Oil Pollution Act 1971 relating to monitoring of possible effects of the discharge of effluent from the Flotta terminal on marine life in the environs of the outfall pipe is stipulated that

'the Company shall, at least once in every period of 12 months calculated from the date of this Exemption, at its own cost carry out tests, of a type and in a manner acceptable to the Secretary of State, to ascertain the state of the fauna and flora of the waters within a radius of 2 kilometres from the said outfall pipe'.

The requirement 'to ascertain the state of the fauna and flora' was clearly a deliberately wide request, requiring consultation with the DAFS Marine Laboratory, Aberdeen.

General biological appraisal. Once each year a general assessment is made of the biota around the effluent discharge site, i.e. in the Hoxa–Switha Sound region. The approach is one of 'skilled eye' assessment to identify any obvious changes. It concentrates on subtidal communities and relates back to pre-operational observations undertaken by IOE, and published data from the DAFS benthic survey when available. Since Orkney Islands Council has extended the commission of Dundee University to undertake monitoring studies, intertidal populations in the vicinity of the discharge site will be monitored under their program and reported annually.

Native shellfish pollutant assessment. Shellfish, mainly scallop (*Pecten* sp.) from the vicinity of the discharge site (i.e. the Hoxa–Switha Sound environs) is sampled at least three times each year and analysed for hydrocarbons, heavy metals and residual chemicals from biocides. Initially, use was made of an experimental seabed test cage system in a natural scallop bed region.

Major attention is given to hydrocarbon analysis, with assessment of at least three fractions, 'aliphatic' hydrocarbons, 'aromatic' hydrocarbons and 'native lipid' fraction. Detailed attention is given to polynuclear aromatic hydrocarbons, particularly those believed to be carcinogenic. A range of analytical systems is employed, mainly gas chromatography, mass spectrometry, and high-pressure liquid chromatography. Since there is risk of heavy metal contamination from polluted ballast water, a limited heavy metal monitoring study concentrating on about four metals is being undertaken. Limited checks are being made for any other chemicals retained in the discharge site, e.g. biocide components.

Caged mussel pollutant assessment. Samples of introduced mussels obtained from a control site (unpolluted), agreed in collaboration with DAFS, held in test cages in the vicinity of the ballast effluent outfall are assayed at intervals for possible contamination. Pollutant analyses with the test mussels will be identical to those for the native scallop assessment. The cage test system required the establishment of a permanent mooring in the vicinity of the discharge site.

Ballast experimental studies at the Flotta terminal. Although not required under exemption 11 of the Prevention of Oil Pollution Act 1971, limited studies are being undertaken on ballast effluent prior to discharge, in order to screen the effluent for specific pollutants and their possible physical and chemical interaction in the oil: water separation plant, and to assess potential biological effects to effluent and effluent components, notably their possible toxicity. Test facilities have been established adjacent to the ballast treatment plant, and include

   (i) effluent continuous and discontinuous sampling facilities;
  (ii) seawater supply/storage;
 (iii) tank test systems (flow-through and recycle);
 (iv) limited laboratory back-up.

There is close collaboration between IOE and the terminal staff, notably the terminal chemist and the ballast treatment process engineers.

THE SULLOM VOE TERMINAL

Comparisons between the two terminals have recently been reviewed in detail. However, several brief but key points are notable. At Sullom Voe

(i) oil throughput and effluent discharge are both considerably larger than at Flotta;

(ii) although British Petroleum is terminal operator, it acts on behalf

Table 1.   Monitoring and other scientific programs in Sullom Voe and Shetland (from ref. 11)

SOTEAG

Chemical monitoring	Hydrocarbons (SMBA [a] Newcastle University)
	Heavy metals (SMBA [b], Strathclyde University)
Biological monitoring	Macrobenthos (OPRU[b])
	Rocky shores (OPRU)
	Soft shores (Dundee University)
	Salt marshes (Imperial College)
Atmospheric monitoring	Lichens (British Museum (Natural History), BP)
Upper basin	Sediment, water column interactions (SMBA)
Seabird monitoring	(Aberdeen University)

Terminal operator

Monitoring at effluent diffuser	(IOE, Heriot-Watt University)

DAFS
Hydrography
Fisheries effects
Macrobenthos production

NERC
Monitoring of Shetland seal stocks (SMRU [c])

NERC/DoE
Biological effects monitoring (IMER [d])

[a] Scottish Marine Biological Association (Natural Environment Research Council).
[b] Oil Pollution Research Unit.
[c] Natural Environment Research Council Sea Mammal Research Unit.
[d] Natural Environment Research Council Institute for Marine Environmental Research.

of a large consortium of offshore operators covering all the fields using the Brent and Ninian pipelines;

(iii) management of all terminal operations is much more complex, with Shetland Islands Council intimately involved;

(iv) the original environmental program (impact and pre-operational) was the responsibility of the Sullom Voe Environmental Advisory Group and the present monitoring program is governed by the Monitoring Committee of Shetland Oil Terminal Environmental Advisory Group (SOTEAG) which in turn is funded by the Sullom Voe Association Ltd.

Much of the extensive environmental program undertaken at Sullom Voe is non-statutory coordinated through SOTEAG, although statutory studies are undertaken, as at Flotta, by DAFS and by a contractor (IOE) for the terminal operator under exemption terms of the Prevention of Oil Pollution Act 1971. This latter monitoring program by the operator has also been interpreted to cover the consent terms for effluent discharge under the Zetland County Council Act 1974.

The basis of the monitoring and related scientific programs in the Sullom Voe environs (Table 1) with some general views on monitoring strategy was discussed recently by Foxton.[11]

REFERENCES

1. Johnston, C. S. (1980). Sources of hydrocarbons in the marine environment. In *Oily water discharges—regulatory, technical and scientific considerations*, ed. C. S. Johnston & R. J. Morris. Applied Science Publishers, London, UK, pp. 41–62.
2. Johnston, C. S. (1976). The design of onshore oil/water treatment systems—environmental and technical criteria. In *The separation of oil from water for North Sea operations.* Institute of Offshore Engineering seminar, Heriot-Watt University, Edinburgh, UK.
3. Johnston, C. S. & Halliwell, A. R. (1976). Environmental considerations in the design of ballast water outfalls. In *OTC* 2446, vol. 1, p. 235.
4. Johnston, C. S. & Morris, R. J. (eds.) (1980). *Oily water discharges—regulatory, technical and scientific considerations.* Applied Science Publishers, London, UK, 225 pp.
5. Moor, I. R. (1980). Oil and water separation—oil loading terminal, Flotta. In *Oily water discharges—regulatory, technical and scientific considerations*, ed. C. S. Johnston & R. J. Morris. Applied Science Publishers, London, UK, pp. 107–18.
6. Trainor, R. W., Scott, J. R. & Cairns, W. J. (1976). Design and construction

of a marine terminal for North Sea oil in Orkney, Scotland. In *OTC* 2712, vol. 3, p. 1067.

7. Halliwell, A. R., Johnston, C. S. & Ramshaw, R. (1974). *The effect on the marine environment of discharges into the sea from the Flotta project, Orkney.* For Bechtel International Ltd on behalf of Occidental of Britain Inc., May 1974.

8. Jones, A. M. & Stewart, W. D. P. (1974). *An environmental assessment of Scapa Flow with special reference to oil developments.* Department of Biological Sciences, University of Dundee, UK.

9. Cairns, W. J. and Partners (1975). *Flotta, Orkney, oil handling terminal, Report no. 3—marine environmental protection.* Occidental of Britain Inc. and W. J. Cairns and Partners, Edinburgh, UK, for Associated Companies.

10. Jones, A. M. (1974). The marine environment of Orkney. In *The natural environment of Orkney.* Nature Conservancy Council symposium, Edinburgh, UK, No. 1974.

11. Foxton, P. (1980). Perspectives in biological monitoring. In *Oily water discharges—regulatory, technical and scientific considerations,* Ed. C. S. Johnston & R. J. Morris. Applied Science Publishers, London, UK, pp. 199–208.

24

Protection of the Marine Environment of the Norwegian Continental Shelf

R. MARSTRANDER

State Pollution Control Authority, Oslo, Norway

ABSTRACT

On the Norwegian continental shelf oil- and gas-related activities potentially harmful to the marine environment are regulated in accordance with the Paris Convention. Exploration and production result in a continuous flow of hydrocarbons and chemicals into the sea. The effects of spillwater and, at Statfjord, contaminated mud, are probably negligible. Accidents possibly involving 15 000 tonnes of oil per day are the greatest threat, particularly to fisheries and rich seabird life. Oil spills are unlikely to permanently affect sea life, although there may be long-term consequences to the littoral. Spills can be prevented or their consequences mitigated by prevention, confinement of spilled oil near the platform, mechanical recovery and use of dispersants. Prevention is the most important, mainly for economic reasons. The State Pollution Control Authority and the Petroleum Directorate now cooperate on stricter environmental controls on North Sea operators. Considerable attention has been paid to development of spill emergency measures by operators and coastal municipalities, but available recovery equipment will work only in waves up to approximately 3 m high and in currents of up to 1·5 knots. Use of chemical dispersants has been restricted.

INTRODUCTION

Threats to the marine environment on the Norwegian continental shelf include commercial fishing, the merchant fleet and activity related to the development of offshore oil and gas resources. Petroleum exploration and production result in a more or less continuous flow of hydrocarbons and other chemicals into the sea. Oil spills up to an estimated maximum of 15 000 tonnes per day are possible.

The discharge of hydrocarbons from oil-related activities on the Norwegian continental shelf is regulated in accordance with recommendations for the North Sea under the Paris Convention.[1] The effects of discharges of chemicals, waste water and at Statfjord cuttings containing hydrocarbons from the oil-based mud are probably negligible and are not commented upon further here. Accidents related to oil and gas activities are the main threat. The huge fisheries on the northern part of the shelf and rich seabird life along the coast are particularly vulnerable.

THREATENED ELEMENTS

The coastal areas and the Norwegian continental shelf north of 62° N include spawning grounds for the most important commercial fish species in the north-east Atlantic Ocean. Location of the spawning grounds is ultimately connected with the pattern of currents along the Norwegian coast (Fig. 1) and rich nutrient content vital for fish larvae. The total catch of fish in Norwegian waters is about 10% of the world catch.

The coast of Norway has also a very rich seabird life and a quantitative description is now being prepared by the Directorate of Wildlife and Fisheries.

RISKS FROM OIL SPILLS

Assessment of the risks arising from oil-related activities involves

    (i) assessing probabilities of accidents which include oil spill;
    (ii) drift patterns of oil spills;
    (iii) weathering effects on oil;
    (iv) effects of the oil on the marine environment, seabirds, etc.

The majority of the oil spills along the coast and in harbors are small and are normally caused by technical failures, overfilling, or illegal discharge from ships. Such spills are kept under surveillance, and they will always be dealt with if they are a danger to recreational areas, bird colonies, fish breeding and so on.

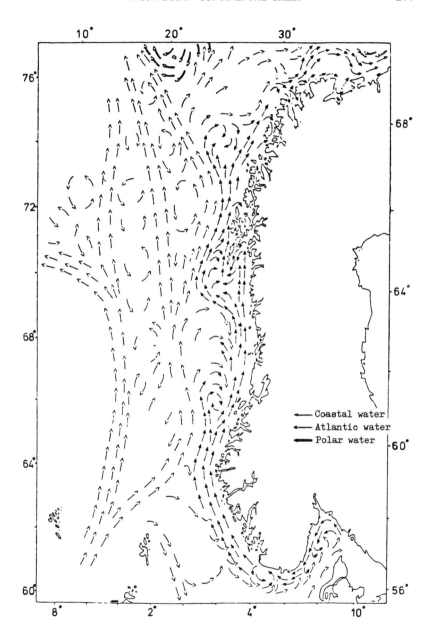

Fig. 1. Ocean currents along the Norwegian coast.

About two to four times a year, however, more serious discharges can be expected from, for example, leakage of bunker oil from wrecked ships, damage to storage tanks or tankers running aground. The largest pollution threats are from crude oil carriers and other offshore activities.

It is difficult to assess the risk from tanker traffic in narrow Norwegian seaways and fjords, and to be confident about the causes and magnitudes of accidents. Several scientific analyses indicate that the possibility of a total tanker casualty is very small. More likely is partial wreckage, with the rupture of a limited number of tanks. A spill of up to 30 000 tonnes close to the coast or in narrow and sheltered waters might be expected from this type of accident.

Offshore oil spills can vary from very small to large and catastrophic and may be caused by operational errors, tank or pipe ruptures or blowouts. While the small spills normally dissolve in sea water without significant damage before reaching the shore, medium and large spills need active clean-up. In rare catastrophic cases, discharges of 12 000–15 000 tonnes per day over several months may be anticipated. Conceivably a blow-out with fire could wreck a platform in such a way that one or several additional wells could blow and further increase the spillage.

Any blow-out in the North Sea must be considered a serious case with potentially large-scale damage to marine life and shorelines. Maximum efforts by the operator and the state will therefore be needed in order to limit the effects of the spill.

The probability of an accident with oil spill as a consequence of drilling a single well or any other single activity is small. But given the level of activity in the North Sea today there is a statistical probability of a blowout every 5–10 years and a major catastrophe every 50 years.

Drift patterns from oilfields on the Norwegian continental shelf have been studied. Numerical simulation models have yielded useful data offshore but have been of less value in coastal waters. The models generally indicate that less than 10% of an offshore oil spill will reach the shore because of weathering effects, biodegradation, dissolution and dispersion.

It is generally judged that an oil slick will not have toxic effects on the marine environment after 15 days. Based on this assumption, calculations have been made related to herring and a possible accident on Haltenbanken for the years 1973, 1975 and 1976. In one year a maximum of one-third of the larvae would be killed (Fig. 2). In the two other years no larvae would be killed.

To better understand both the degradation of oil in sea water and the

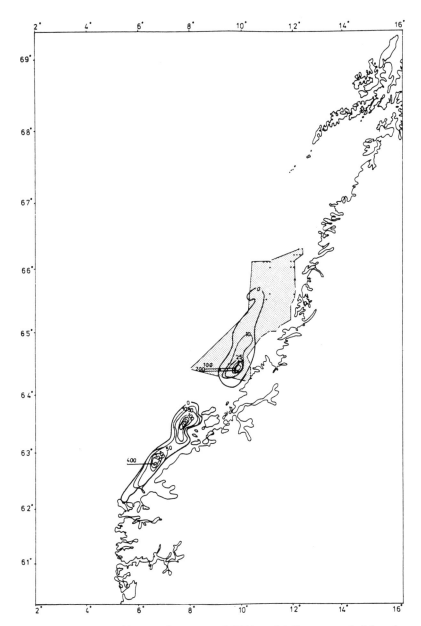

Fig. 2. Distribution of herring larvae, April 1976, and drift pattern of oil less than 15 days old at Haltenbanken, off Norway. Stippled area indicates drift pattern of oil less than 15 days old. Contours represent numbers of larvae per square metre of surface.

effects of oil on the environment, a marine pollution research and monitoring program has been started, financed by the Norwegian government,
which will last until 1983–84. Studies carried out by that program and by
the Norwegian Institute of Marine Research indicate that

(i) an oil spill is unlikely to have a lasting influence on pelagic marine
life;
(ii) under specific circumstances there may be a reduction in the total
number of fish from a given year's stock;
(iii) biological consequences to the shoreline may be long lasting.

PREVENTIVE SAFETY

Although the safety of ships and regulation of marine traffic is an international issue this section concerns only the North Sea.

It has frequently been stated in Norway that highest priority must be
given to increasing safety of North Sea oil operations. This includes technology as well as training of personnel and deals with all aspects of safety,
i.e. human lives, environment, and economy. Human lives and economy
dominate, whilst environment comes last. Because of the need for safety
of personnel working offshore and because the economic consequences
of an accident are of such magnitude, operators tend to do whatever is
possible to increase safety in any case. They argue that it is not possible
to do more to protect the environment than is already done for economic
reasons.

In economic terms, environmental considerations take low priority
because of the way different investments in the oil industry are valued.
Calculations are based on discounted values and a high internal rate of
return, which means that damage to the environment from any future
event is given a small value at the time of valuation.

In the past, environmental specialists and authorities such as the State
Pollution Control Authority (SPCA) were not involved in decisions with
regard to safety offshore. Now, however, the SPCA and the Petroleum
Directorate cooperate to see how environmental measures can be more
strictly included in requirements. This requires consideration of both the
conceptual stages of a development and of the technical requirements
when methodological concepts have been decided upon.

DEVELOPMENT OF ENVIRONMENTALLY ACCEPTABLE
METHODS

Even if everything possible is done to increase safety there still is a chance of a situation getting out of control. In the marine environment the amount of oil spilled per unit time and in total are important. On site, the first concern is human life and thereafter economic consequences.

To get an offshore blow-out under control a series of measures can be taken of which the last and most drastic is a relief well. Relief wells can take up to 100 days to drill, during which an enormous amount of oil may be released. It is therefore important to consider other ways of reducing the amount of oil spilled as a result of an accident.

During and immediately after large accidents like those on Bravo and Ixtoc I many ideas are presented on how to capture spilled oil at the well, e.g. the Mexican 'Sombrero' and different damage systems on North Sea platforms. None has so far proved practicable. An economically feasible solution is obviously difficult to find.

The production wells at Ekofisk and Statfjord illustrate the economic aspect. At Ekofisk the diameter in the well pipes is about 10 cm which with the pressures available gives a possible accident rate of 4000 tonnes per day. At Statfjord the pipes are about 20 cm and accident rates are in the range of 15 000 tonnes per day. These differences are clearly for technical, economical and geological reasons. For the well-being of the marine environment the Ekofisk production rate is obviously to be preferred to that of Statfjord.

Mechanical recovery of oil from the sea surface has limitations (see below).Because of this there is a clear need for more environmentally sound methods for the development of offshore oil- and gasfields than those we have today, methods that will reduce the amount of oil spilled in an accident. There are reasons to be optimistic that such methods may be developed; new ideas are coming from various sources and there is a growing understanding of environmental needs.

There are two reasons to believe that the methods now in use will be changed. First is that development of fields at greater depths requires new methods of approach, e.g. on the seabed. The second is purely economic. A recent study estimates costs of developing oilfields on the Norwegian continental shelf. The total cost for Ekofisk II, III, IV and pipelines, Statfjord A and Frigg I and II is over 7×10^9 (Table 1).

It seems likely that methods which use approximately 50% of the costs

Table 1. Cost of developing oilfields Ekofisk II, III, IV, and pipelines, Statfjord A, Frigg I and II

Phase	$US (millions)	%
Engineering and management	697	9
Construction	2 018	28
Equipment	563	8
Transport	1 948	27
Hook-up	1 527	21
Insurance	131	2
Miscellaneous	377	5
Total cost	7 261	100

on hook-up and transport will have to be changed. With the environmental considerations more prominent today than in the past, it is reasonable to assume that future methods for development of oilfields will be more acceptable environmentally.

MECHANICAL OIL RECOVERY OF SPILLS

Because accidents are to some degree inevitable, different means for mechanical oil recovery of spills have been developed. The protection offered by recovery equipment can never be complete and is likely to be far from that under North Sea conditions. Mechanical oil recovery equipment has to be operated from boats and by trained personnel.

Norway has a very long coastline and not many accidents will normally occur. This means oil recovery organisation is based on boats and men normally occupied with other tasks. Only the recovery equipment itself and a relatively small number of men will be permanently employed for oil recovery. Norway therefore has three mutually complementary kinds of organisation for oil recovery:

 (i) offshore operators;
 (ii) the state, through SPCA;
 (iii) coastal municipalities.

Since none of the three is exclusively employed for oil recovery except for a few key people, exercises and a written framework of contingency plans are of the utmost importance. Since it cannot be predicted where or in which way an accident will hit, a certain degree of improvisation is inevitable. The points mentioned above are very important when considering the organisation. Organisational difficulties and even malfunctions are bound to reduce the effectiveness of large oil recovery operations.

Oil Recovery Equipment

In Norway, operators on the Norwegian continental shelf and central and local authorities along the coast are responsible for keeping oil recovery equipment ready for use in case of accident. Local authorities have light equipment for use in narrow waters and against relatively small amounts of oil.

The operators, who are responsible for oil recovery on the continental shelf, have eight skimmers and 3000 m of heavy ocean oil booms south of 62° N. The latter are expected to function in 2·5 m significant wave height (swh) and currents less than 1·5 knots. North of 62° N, the operators have ten skimmers and 4000 m of heavy ocean oil booms expected to function in 3·0 m swh and currents less than 1·5 knots. In addition, operators will use boats already in use for daily operations and fishing vessels.

The SPCA has the following equipment at each of 12 depots along the coast:

  800 m heavy coastal oil boom
  1000 m medium size oil boom
  three skimmers of different sizes
  one boat

At some depots there are fenders and other equipment for use when emptying ships of oil. In addition 30 boats from the ocean-going fishing fleet are hired to take part in oil recovery operations.

All the above recovery equipment is based on the principle of booms and skimmers. The opinion of various experts is that above 4·0 m swh so much oil will be dissolved or dispersed by wave action that no surface oil will be left for the oil booms. It is also estimated that by 3·0 m swh the efficiency of oil booms will be significantly reduced. Depending on the time of year and the location, waves less than 2·5 m swh are probable for 30–77% of the year. Similarly, waves of less than 3 m swh are probable for 48–87% of the year and waves of less than 4 m swh for 70–95% of the year (Table 2). Operating enough equipment to handle a blow-out of 10 000–20 000 tonnes per day requires 30–40 boats to stretch booms, transport oil, etc. In addition to physical limitations oil recovery equip-

Table 2. **Calculated wave heights for proportions of periods at different locations on the Norwegian continental shelf, 1974–76**

| Location | Given wave height (m) | Percentage of period below which swh^a is less than given wave height | | |
		January	July	Yearly
Famita (Ekofisk)	1.5	24	39	29
	2·5	49	70	59
	3·0	61	81	71
	4·0	80	93	87
Kråkenes (Haltenbanken)	1·5	23	45	32
	2·5	49	75	62
	3·0	60	85	74
	4·0	76	95	86
Polarfront (Statfjord/	1·5	16	31	22
Haltenbanken)	2·5	36	61	47
	3·0	48	73	60
	4·0	70	90	79
Ami (Tromsøflaket)	1·5	30	52	40
	2·5	59	77	69
	3·0	70	87	80
	4·0	84	95	91

^{a}swh = significant wave height.
Source: Parliamentary Report No. 65 (1977–8), Oslo, Norway.

ment may also be difficult to use in darkness. During winter periodic icing is also likely.

The above represents traditional oil recovery equipment, although its size is exceptional. During the last few years, however, a series of new concepts for oil recovery have emerged. Roughly, these can be grouped as follows:

(i) oil trawl moved by two supply boats;
(ii) a boat with booms and skimmers rigged out from both sides of the boat;
(iii) an oil collecting catamaran with a speed of 5 knots;
(iv) a system based on absorption of oil from the surface.

All of these systems are under study, some undergoing practical experiments, but none is yet operational. The new systems are subject to the same physical limit of 4 m swh but they are easier to operate and therefore more efficient.

CONCLUSIONS

Even if studies show that oil is not as toxic or such a threat to marine life as was thought in Norway some years ago, it is still a threat to fish larvae and birds, and in general represents a very dirty kind of pollution. Besides, oil has become a valuable resource.

There are mechanical limitations in oil recovery, but for economic reasons oil operators do their utmost to prevent accidents. So far there has been little real conceptual work oriented towards more environmentally acceptable methods for development of offshore oilfields, but improvements are likely. Certainly if methods are not improved, every now and then large-scale pollution from accidents is inevitable because climatic conditions severely restrict the use of recovery equipment in the North Sea.

REFERENCE

1. *Petroleum exploration north of 62° N.* Parliamentary Report No. 91 (1975–76), Oslo, Norway.

BIBLIOGRAPHY

Petroleum exploration north of 62° N. Parliamentary Report No. 57 (1978–79), Oslo, Norway.

The uncontrolled blow-out at Ekofisk (the Bravo platform), 22 April 1977. Parliamentary Report No. 65 (1977–78), Oslo, Norway.

Petroleum finds north of 62° N. Possibilities and consequences. Norges Offentlige Utredninger, 1980: 25.

Blow-out on the Norwegian continental shelf. Norges Offentlige Utredninger, 1979: 8.

Summary record from the first meeting of the Paris Commission, The Hague, 27–30 November 1978. Secretary of the Paris Commission, 48 Carey Street, London, UK.

The effects of oil pollution on the marine environment north of 62° N. Norwegian Marine Pollution Research and Monitoring Programme, Report No. 1, (1979), Oslo. Norway.

Andunson, T. *et al., Fate of oil spills on the Norwegian continental shelf.* Norwegian Continental Shelf Institute, Trondheim, Norway.

Oil spill in Norwegian waters. Det norske Veritas. Report 79–1030. Hovik, Norway.

Description of the climate on the Norwegian continental shelf. Scientific Report No. 18. Norwegian Metereological Institute, Oslo, Norway.

Cost study Norwegian continental shelf. Part 1. Recapitulation of the development, evaluations and recommendations. Report dated 29 April 1980 of a commission appointed by the Norwegian government. Statens trykksakekspedisjon, Oslo, Norway.

Discussion

P. Daniel (Department of Architecture, University of Edinburgh, UK). Could Dr Norcross explain how the Wilhelmshaven plants are to be landscaped?

G. Norcross. The old sea dyke, some 10–15 m high, provides screening from the land side. Adjacent to it, artificial sand dunes with trees and shrubs will provide screening from another direction so that the ICI plants will, with the exception of an occasional chimney, be unobstrusive from the villages (1–2 km distant).

P. Daniel. How willingly did the local population accept such an enormous building complex?

G. Norcross. In general they accepted it, but there are vociferous action groups in that part of Germany and some objections continue. However, after the early discussions the authorities took on the job of justifying the project to the locals. With the help of the Land government local authorities had spent millions of Deutschmarks raising the land level by as much as 5 m from tidal muds to become industrial sites and so have a strong incentive to see the land successfully developed. Helped by ICI and its consultants, all legal objections have been overcome, although a few citizens will never be satisfied.

H. J. Moody (Tayside Regional Council, UK). Dr. Norcross mentioned that some 21 plants were planned in the long-term but their exact form or use is not yet known. How were these evaluated and explained to the general public? How would Dr Slater undertake hazard assessment for 21 unknowns of that kind? This is of relevance in Scotland where we tend to identify strategic industrial development sites prior to particular plans. Technological change is rapid, particularly in the chemical industry, so the public might unknowingly consent to a most unwelcome development.

G. Norcross. ICI had to consider very carefully the kind of plants it might build in the complex and, through experience of similar plants elsewhere, to estimate probable impacts, e.g. of noise, water pollution, etc. The ICI Brixham Laboratory is closely consulted in this sort of work. If there is a very high probability, for example, that the impacts will be within expected limits, then it is worth going ahead. Obviously potential impacts must be very carefully considered because if only three or four plants could be built on such a big site it would be uneconomic.

D. H. Slater. In hazard analysis (which Cremer & Warner did for Wilhelmshaven) assumptions have to be made, even if hypothetical, and a complete picture of all risks built up. The assessment has to be publicly displayed and any challenges answered.

Anon. Hazards are assessed, including contour maps of risk, on the basis of assumptions from a firm not involved in works operation and possibly with no continuing responsibility. In the UK, a different procedure may be used, e.g. in underground mining the mine manager is personally responsible. There is similar individual responsibility for every dam and reservoir over a certain size in the UK. On nuclear installations there is a government department but apparently nothing similar in the petrochemical industry, where a contractor assesses risk. The contractor's analysis may be correct but has no statutory backing.

D. H. Slater. Mining may not be a good example because of accident rates, etc. Separation of production and safety responsibilities is important, because the man involved in both is under tremendous pressure. The job of the consultant is much like that of an independent auditor of accounts. While you have perfectly competent accountants in-house, the external auditor provides a more objective independent check.

   If assumptions are made in assessing a hazard then they must be stated. Local authorities can then incorporate them in planning conditions.

G. Beveridge. In addition to good design, a plant must be well operated. Management in the chemical industry is very important. Good experience is far more valuable than licensing of chemical engineers.

G. Norcross. In Britain we have the Health and Safety Executive. In Germany a safety engineer must be employed by the chemical firm and is answerable to the authorities for safety matters. After Cremer & Warner had done their hazard analysis of the Wilhelmshaven complex it was assessed by a Gutachter (expert consultant) reporting to the regional administration. Various Gutachter assess different aspects of the proposals.

D. C. Monk (BP Petroleum Development, Shetland Islands, UK). The exemption certificates at Sullom Voe state an average of 10–15 ppm of oil in water, with a maximum of 25 ppm. Values in excess of 15 ppm are permitted only 20% of the time.

At Sullom Voe up to 6000 m³ per hour of effluent less saline than sea water may be discharged in future. Apart from any oil-induced effects there may well be some degree of environmental change. How far can one go from the point of discharge before that change becomes unacceptable?

C. S. Johnston. A limit figure on an exemption certificate is of limited relevance. We should not be obsessed with such figures unless, at least, we know the basis and environmental significance of the measurement, but include a dilution factor in the consent to discharge formula.

Fresh water discharges do tend to create mini-estuarine systems, and the changes would be acceptable if the fresh water source were a natural stream. However, we should question the reason for monitoring any contaminants, whether oil, heavy metals or others. Monitoring can also help the operator to check if his treatment plant is working properly.

Anon. We should distinguish between carcinogenic and assimilable contaminants. Dilution is adequate for the latter but not the former.

C. S. Johnston. Perhaps substances should be categorised, a measure of risk put on them and the result incorporated into the formula for the protection of the environment. A further factor which should be considered is the physical nature of the water receiving effluents, not just agitation but also particulate loading, which affects absorption properties.

Anon. In some cases it may be best to apply limiting figures at the pipe rather than further away where levels may be below our technical ability to measure them.

C. S. Johnston. If we have techniques or can foresee having techniques to reduce emission levels we should consider them. However, a low level of one contaminant may not indicate low levels of some other because a given technique is not equally efficient for all contaminants. A great many variables are involved and in establishing precise emission standards we are in danger of overlooking reality.

J. M. Baker (Field Studies Council, Pembroke, UK; also representing the International Union for the Conservation of Nature and Natural Resources, Gland, Switzerland). Bioaccumulation can occur within a single organism or up food chains. What evidence is there for bioaccumulation of hydrocarbons up food chains?

C. S. Johnston. Undoubtedly both do occur. Bioaccumulation and depuration are in a grey area requiring more work. All work on depuration, or the lack of it, really comes from a single laboratory in the world, and that mainly concerns low molecular weight aromatic compounds.

Section V

ISSUES, LESSONS AND CONCLUSIONS

Chairman

Dr John H. Burnett, FRSE, Principal and Vice-Chancellor, University of Edinburgh, UK

25
Summing Up — The UK View

P. JOHNSON-MARSHALL

*Department of Urban Design and Regional Planning, University of
Edinburgh, UK*

INTRODUCTION

This conference was initiated when a group of environmentalists met
on an entirely voluntary basis to consider Scotland's particular oil and
environment experiences. It was decided that an international conference
would be a timely and appropriate forum to focus on North Sea problems,
particularly for Scotland and Norway. The conference would link with
the European community and attract contributions from across the
Atlantic.

The objectives of the conference were to examine the importance of
oil in the community and over-dependence on one energy source, and to
examine its impact on the human environment in the light of up-to-date
information and the experience of those actually involved.

In particular, the idea was to have a confrontation, albeit a friendly
one, between ecologists and environmentalists on one hand, and oil
development experts on the other. Scotland is an appropriate place for
such a confrontation, because a similar experience happened here during
the 19th century, and we are left with much of the physical legacy.
Scotland is relatively well prepared for the second industrial revolution,
through the Scottish Development Department, the Town and Country
Planning Acts, the Highlands and Islands Development Board, the
Nature Conservancy Council, and the Countryside Commission. Such
a confrontation would be timely as oil developers respond to a kind of
technological imperative based on public demand, which is so clearly

expressed at every petrol pump that no politician dare whisper a word against oil.

CONFERENCE ACHIEVEMENTS

The environmental context for the conference was set by *Larminie*† and *Nicholson*. *Larminie* was, of course, greatly influenced by the technological imperative coupled with the need for engineering competence, but was also concerned with social and economic implications. He emphasised that the oil world was a very different world from that of most people. *Nicholson*, in a broad philosophical contribution, emphasised that round table negotiation in a proper environmental context is now essential, and is in general accepted by all parties concerned.

The control of development falls largely to planners. *Lyddon* covered the sophisticated planning methods used at national and local levels in Scotland, the use of impact analysis as a complement to the statutory planning system, and the democratic procedures built into it. While *Gjerde* provided the Norwegian perspective on planning, *McCarthy* reminded us of the complexity of biological aspects of planning. Many problems range far beyond normal statutory governmental procedures. He brought out the difficulty of assessing complex information to enable clear decisions to be reached quickly, and noted that outline planning permission tends to commit authorities at perhaps too early a stage.

A basically new technological installation of a major kind was described by *Carder*, delivering *Taylor's* paper on Mossmorran. While emphasising the importance of public relations he perhaps failed to note that some local citizens still feel there are environmental hazards involved. However, the extraordinary hazards which oil men themselves have to take north of the 62nd parallel was described well by *Lind*.

Stuffmann showed convincingly that the EEC is not just a place where they argue about fish and agriculture, but where knowledge and policies on environmental assessment are developed as well. There is hope that Europe may yet be one place with a common policy for oil and environment, so let us hope that Europe can respond to the philosophical pleas so dynamically expressed by *McHarg*.

Dean and *Graham* as a team described the invisible world of pipelines

---

† Names in italic type are those of contributors to these proceedings.

and cables underneath Britain, which has been built with careful environmental planning. Landscapes cannot be preserved in a static way, but their evolution can be carefully planned. Both *Sargent* and *Fehily* showed that good new landscapes can indeed be planned and made, through careful attention to the right environmental factors.

In the USA one of the most powerful influences on the character of the environment is the country's consuming demand for energy. *Harrison* explained that political concepts of liberty and freedom make planning difficult to achieve in the USA, but nevertheless the challenge of air and water pollution and other health hazards was being taken most seriously in a very large and complex land. Perhaps the Americans may glean some lessons from the experience of the small and delicate environment of Shetland, whose solutions to physical, biological, social and economic problems were so well outlined by *Fenwick*.

CONCLUSIONS

It would be naive to expect clear-cut conclusions from the enormously complex subjects discussed, but there is clearly value in having a group of disinterested persons, as the organising committee of the conference were, concerned about real problems. A confrontation of this kind between oil men and environmentalists develops mutual understanding and respect, and each group learns from the other.

It is perhaps worth noting that Britain has a planning system designed for local needs and democracy, but which in general has been able to deal with the vast new international scale of oil-related development. The drama of new technology at a large scale has been clearly demonstrated, but the Scottish environment in particular has been able to come to terms with it, and even the great multi-national oil companies are now willing to talk around the table.

All the problems are not yet solved, but many lessons are being learnt, particularly the need for good monitoring and for more broadly based expert teams, but looking to the future we must also consider what to do with the major installations when the oil and gas are exhausted.

26
Summing Up—The Norwegian View

K. STENSTADVOLD

The Institute of Industrial Economics, Bergen, Norway

In Norway, environmental impact assessments have emphasised the economic and social effects of oil-related development and paid relatively little attention to natural, physical and biological environments. There may be many reasons but two main causes should be stressed.

The first is the geography of Norway. Norway has a long and sparsely populated coastline. Even the second largest city of Norway (Bergen, population about 200 000) does not have a complete sewage treatment system, yet one may swim only 3 miles from the city center. There has been some improvement in sewage treatment recently, but tradition alone might dictate otherwise. Because of the small population, nature has been able to absorb the impact. In addition Norway has little large-scale industry. With the exception of urban areas there has been, until recently, little need for constant and systematic physical and land-use planning. Even before the advent of offshore oil Britain was, of course, in a completely different situation.

The second reason has historical roots and concerns certain cultural values and a tradition of small-scale society in rural areas. As oil has come to Norway this has led to a very keen interest in how it will influence the way of life in remote areas. For example, there is concern about how oil-related developments will influence traditional industries, such as agriculture and fisheries. There is also interest in how oil-related industry could be adapted to suit small communities.

One might add a third reason, which relates to the division of responsibilities between the different levels of government in Norway. In the UK both physical and socio-economic environmental assessments are the responsibility of local government. In Norway there are not yet any clear responsibilities but a local authority is mainly responsible for socio-economic and land-use planning, while pollution and similar types of physical impact are the concern of the State Pollution Control Authority (SPCA, *Marstrander*†). A local community may comment upon any proposals for development but there is surprisingly little local debate on such issues. In general the local community is convinced that the SPCA will take care of pollution matters.

An example is the recent enquiries into landing Statfjord gas in Norway. The most voluminous reports concerned labor markets, local industry, local authority finance, development of infrastructure, economic multipliers, etc. Little space was devoted to physical impact, pollution and so on.

A significant difference between the approach in Britain and Norway, *Gjerde* and *Lind* showed, is a marked tendency to start with a reference to national oil policy even when discussing local pollution matters. The argument is that a high level of offshore oil activity carries a correspondingly high level of pollution risk, and the same applies to the rate of change in local communities affected by oil. This, of course, at times leads to very general discussion of oil-related impacts, but it makes everybody involved aware of the complexities of oil-related development.

That argument becomes even more pronounced when the areas north of 62°N were to be explored. There is only a handful of towns north of Trondheim with populations over 10 000, and it was felt that the traditional way of developing petrochemical industry would not be suitable as a general regional development tool. A study was therefore commissioned to investigate the possibilities and limitations of subdividing petrochemical plants into small units in order to better suit them to local socio-economic conditions.[1]

While Norway has shown increased awareness of social issues she is not necessarily prepared in the best way. For example, Norway tends to start from scratch for each new landfall decision. The first study started in 1973 for landing gas from Frigg at Kårmoy. Then in 1976 was a study for landing oil from Statfjord at Sotra and for gas from Statfjord in 1980.

Norway could learn from Britain's experiences, e.g. Scotland's survey of the entire coastline for conservation zones and preferred development zones (see p. 25). One reason Norway lags behind may be a feeling

---

† Names in italic type are those of contributors to these proceedings.

that Norway has enough area and sites to choose between; another may be that Norway does not have the equivalent of the Nature Conservancy Council. In Norway the nature conservation lobby is private and is mainly concerned about hydro-power development. The main opposition on physical and environmental grounds has come from agriculture. Will Norway take the opportunity to learn from UK experience or must it first have irreplaceable countryside spoiled by a large petrochemical plant? There are other lessons to be learned from both sides of the North Sea. While the physical environment has monitoring as a condition in the planning consent, there is no monitoring of any kind of the socio-economic environment, e.g. of the influence of big projects on local labor markets. Here, Norway may look to Britain. Shetland has been running a good system of labor market monitoring since 1977. Norway may have lagged behind because of the poor reputation of social scientists in the late 1960s and early 1970s. In the recent Statfjord gas study guesses, albeit informed ones, have been made at an unprecedented scale and important topics dropped altogether because there was no basic knowledge or background information.

Another lesson from the present conference is the need for increased access to and use of information amongst the parties involved. In Norway this has improved over the last few years. From the information in various papers (these proceedings) and other sources it seems clear that in those projects where information has flowed freely from the early stages there have been fewer problems, e.g. environmentalist groups need not litigate to obtain information if it is already provided, thus avoiding costly delays.

That leads to the subject of contingency planning. The present problems from rushing the Statfjord gas landfall decision illustrate the value of such planning. The first information was published and planning started in May–June 1980. Local authorities had to process planning applications by the end of 1980 in order for a decision on the landfall to be made in May or June 1981. That would be too much of a rush in any case, but the problems are increased by the lack of any planning around the potential landfall sites. Norway should look to Britain and adapt relevant parts of their coastal strategic planning, but local politicians in Norway must be convinced that mapping of potential sites is not necessarily a promise of large-scale development (see p. 63).

Another issue which has been raised and which should be developed is to consider external impacts early in the developers' engineering and planning decisions. The visual side now is well catered for by architects who have to participate in the development of the whole project. Continuously improved and less polluting techniques are also being built into new plant. However, there has been relatively little consideration of alternative ways to develop projects so that the socio-economic benefits are optimised. This is not simply a question of technology but of organisation in a broad sense, including the introduction of such issues at the appropriate stage of project planning. For example, by modularisation of process plant, or by building entire process lines on barges, the effects of large labor forces in remote areas may be substantially modified.

In principle, Norwegian local authorities are more independent than their British counterparts. On the other hand the government has, since the war, followed a policy of equalisation which has meant centralisation of, for example, pollution standards. Thus the SPCA imposes uniform standards in a way alien to the American system (*Harrison*) where effluent quality levels can even be traded (the Bubble Policy—see p. 218). Such a system would not at present be openly accepted in Norway, where in principle a fishing village in the north should have the same standard of sewage treatment as a town in the relatively crowded south-east. In practise the principle is not always strictly applied but Norway could still use its geography more efficiently.

Although Norwegian local authorities have relatively little leeway in practise, they still have strong political voices to sound. This is also strongly felt in the central ministries and by commercial companies so that local involvement has been fairly extensive. It is significant that communities which seem to have benefitted most from oil in Britain, that is Shetland and Orkney Islands, have been those most actively involved in negotiation and participation of different kinds.

Traditionally bargaining power has been in the developer's favor. They are experienced from previous developments in various locations. The local community, on the other hand, is usually involved for the first time. Theoretical studies of bargaining power and of market power in general show that there must be a reasonable balance of power between the participants, i.e. local authority and developer, if optimal solutions are to be found. 'Power tends to corrupt and absolute power corrupts absolutely'.† Where does the balance lie?

Planners, consultants, administrators and politicians must consider whether impact assessment studies always give the information decision-

---

† Lord Acton. *Historical essays and studies.*

makers really want or need. UK impact studies are usually detailed. In Norway the opposite is usually the case, impact assessments being relatively general. With the increasing demand for impact assessment studies what is really needed should be very carefully considered. The majority of papers presented at the present conference have been by developers, local authorities and consultants. They quite naturally bound to say that a project has generally gone well and are unwilling to admit deficiencies. Are all our experiences of oil-related developments so positive and satisfactory as they would have us believe?

REFERENCE

1. Lind, A. *et al.* (1978). *Desentralisering av petroleumsbasert industri.* (Decentralisation of petroleum based industry). NIBR Arbeidsrapport 22/78. Norwegian Institute of Urban and Regional Research, P.O. Box 15, Grefsen, Oslo 4, Norway.

General Discussion

E. M. Nicholson (author). We have rather taken for granted the capacity to relate the very wide range of data, technical, scientific, economic and so on, to the problems of decision-making. It does appear that in a number of the cases that have been given to us, sufficient data have been improvised in particular places to enable decisions to be taken which appear, at least in the early stages, to be working out fairly well, but we must be more sceptical about that in the longer run. Dr Johnston showed, for example, how biological data had to be broken down in order to see the difference between pollution and contamination, the capacity of the medium. We are at a very elementary stage on structuring the handling of the data in order to make predictions.

Much experience has been gained in different parts of the North Sea, but there are almost no facilities for exchanging the data. The people in Flotta and in Shetland have no way at all of relating their own data to that obtained in Norway, in The Netherlands, and so on. The North Sea is one ecosystem yet we are treating it as if it were a lot of fragmentary ecosystems. There will be a bill to be paid for that unless there is better comparative assessment of data and the methods.

One other aspect which has been insufficiently considered is that we are still at a very early stage in the exploitation of oil. The run-down, when oil production begins to taper off, is probably only about 5 years ahead. This conference has said very little about the problems which will arise when we are not rushing to increase output but are trying to slow down the run-down. In this respect Norwegian policy on the rate of exploitation has been very different from the policy in the UK. Should oil production be spread over a longer period, for example?

Many interesting things have come out of this meeting, but we have far from solved all the problems and may be even unaware of the problems that are going to hit us in the fairly near future.

K. Stenstadvold. Is stretching Norwegian oil production to 100 years at, say, 60 × 10⁶ tonnes per year more detrimental to the environment in the long run than producing twice as much in half the time? Much research is required on long-term effects, and the results are important for planning.

P. Johnson-Marshall. We can never get enough information, but decisions must be taken with what is available. More money and more time is needed.

J. H. Burnett. One study of soil surveys by governments of the world indicated that some 15 times as much data is usually collected as is needed for decision-making. This may not apply to the North Sea and oil, but the important thing is not the amount of information but its relevance now and in the future. This, of course, is a matter of judgement.

A. Jackson (Department of Social Anthropology, University of Edinburgh, UK; representative from the Social Science Research Council, North Sea Oil Panel). A topic perhaps insufficiently discussed here has been that of social impact. At this conference environment has been implicitly defined as the physical environment and the fish, birds and other creatures that live around it and in it, but has tended to exclude the humans that live there and are presumably to benefit from our concern for it. Surely risks to community life, the psychological damage done to individuals, are equally if not more important than physical risks.

An instrumental approach to planning for efficient production of oil with the minimum disruption to the physical environment is misguided because it neglects the effects that the oil industry has upon local communities; it is misleading because it avoids confronting the social implications of its activities; and it is misplaced because it emphasises the wrong issues, that is, physical pollution.

The problems which have been discussed here are important, but particularly to us as human beings. Professor McHarg suggested that anthropological models should be considered in the planning process. The social impact of oil-related activities would be a pertinent topic for a further conference.

The overly optimistic picture often presented here, that there are no real problems that cannot be solved technically, overlooks the issue of how the developments affect the people at risk. This is not a very grave problem because humans are very adaptable, but we seem to be doing very little to ameliorate their position. Conservationists are perhaps the major guilty parties. With tongue in cheek, I suggest that they are distracting the attention of the oil companies and the government away from the real issues, that is the social effect of industrialisation. Research on this subject is being undertaken by the North Sea Oil Panel.[1]

How important does Norway consider the social impacts of oil-related development?

K. Stenstadvold. I agree with Dr Jackson's point of view, but am uncertain which approach is appropriate for Norway. Social studies must avoid the pitfalls of the conservationists. In Norway there is a tendency to regard the small crofting/fishing village as the ideal country community. Then all development is alien and should never take place nearby to change it. There is a need for a better approach; my feeling from my recent work is that we have not really hit directly on what we should do. There are several new studies underway in Norway. One just started by the Norwegian Institute of Urban and Regional Planning in Oslo is attempting to identify important factors in a community which may indicate whether the society will be improved or destroyed when large-scale developments are planned. A committee[2] which I will chair is being established to consider priorities in social science research in relation to oil in Norway. The experience of the SSRC North Sea Oil Panel might be most useful. Studies are also being undertaken in the north of Norway, on

---

[1]Byron, R. & MacFarlane, G. (1980). *Social change in Dunrossness; a Shetland study.* North Sea Oil Panel, Social Science Research Council, University of Glasgow, 2 The Square, Glasgow G12 8QQ, UK.
[2]Appointed October 1980 to report March 1981 on *Societal effects of oil activities in Norway*, sponsored by Norwegian Research Council for Science and the Humanities (NAVF), Sub-council for Research for Societal Planning (RFSP), Oslo, Norway.

the relationship between fisheries and the possible socio-economic impact of oil. [3,4]

There are, of course, many studies on the physical impact of oil on fish, as referred to by Mr Marstrander. Follow-up studies on petrochemical works in south Norway will probably begin soon, although it is embarrassing and surprising that they did not start 3 years ago, when the first large-scale oil plants were established. A comprehensive study was done on the impact of a large aluminium smelter on an isolated community (Årdal) some years ago,[5] some of which is obviously relevant, but one case is not enough.

F. D. Hamilton (Royal Society for the Protection of Birds, Edinburgh, UK). Oil and the environment have not always developed in harmony in the North Sea. The first 5 years of development of North Sea oil were, environmentally speaking, relatively chaotic, with concerned people reacting to the industry. Order was slowly established only when the direction oil development itself was taking became clear. It is also, perhaps, pertinent to note that nearly all oil-related developments have proceeded beyond what was envisaged at the time of the original planning application.

While the environmental efforts made by the oil companies have been entirely creditable it is doubtful if any oil company raises its environmental standards without outside pressure, whether from government or elsewhere. As an example one might ask if the environmental standards of oil companies' activities are the same in Thailand or Venezuela as they are in the North Sea.

J. McCarthy (author). There are important things which should be understood when using the words conservationist, natural history interests, birds and so on. Most mature conservationists are beginning to understand much more than before that something is not just 'for the birds'

---

[3] Marit Reutz, Institute of Fishery Studies, University of Trømso, on *The impact of oil on recruitment to the fishing industry.* Financed for 1980–83 by Norwegian Fisheries Research Council.

[4] Audun Sagberg (project manager), Nordland Regional College, Bodo, on *Coastal communities and oil: changes in economic structure of the Nordland county.* Financed for 1980–82 by Norwegian Fisheries Research Council.

[5] *The Årdal project.* Eight reports 1971–75. Norwegian Institute for Urban and Regional Research, Oslo, Norway.

but that natural history, ecology, conservation are just as much part of our culture as a whole variety of human activities. There is not a separate compartment into which we slot birds, etc. We are beginning to understand that wildlife is indeed part of our culture in the broader sense of the word, remembering that wildlife includes literally the environment in which we are living.

In Scotland, particularly, maintenance of these resources is not just for cultural reasons. The maintenance of fisheries is obviously economically important. Also tourists increasingly come to this part of the world and spend money to see something that they cannot see elsewhere. So there is a clear economic element in conservation.

The NCC is required to be responsive to public interest. The Countryside Act requires it to have regard for socio-economic interests in all its activities. It is for others to judge just how capable NCC is of devolving that responsibility.

NCC is also responsive to a wide range of voluntary organisations. These organisations are most important contact points for official nature conservation organisations to ensure that they are, in turn, responsive to public opinion. An organisation such as the Royal Society for the Protection of Birds (RSPB), one of the fastest growing organisations in Britain, includes within its membership a very wide range of people with interests that go simply beyond looking at birds to a much more general interest in the countryside at large. Nature conservation and wildlife must not be put in some special category because they make their own contribution, with others, to the health and general economic and cultural welfare of the human community.

E. M. Nicholson. Contrary to what Dr Jackson has suggested I would stress that conservation today is not mainly concerned with wild-life, etc. IUCN has recently launched a World Conservation Strategy (see p. 50) in some 32 national capitals. It might better be called a World Conservation and Development Strategy since it is largely devoted to the reconciliation of human aspirations with the limitations of our resources. It is the resource base which is being chopped away with grave consequences to those whose living depends on it and we are not getting nearly enough help from legislation or social services.

The fisheries of the North Sea provide a good example, where large-scale unemployment has arisen because the industry refused to listen to very clear warnings given by conservationists about over-fishing. That great conservationist, Frank Fraser Darling, long ago made one of the broadest ever surveys of human problems of a region in relation to conservation of natural resources and culture.[6] For more than 30 years the Nature Conservancy Council has developed a complete Hebridean island community on the Island of Rhum.[7] How many sociologists have ever set foot there?

It is not helpful to talk as if conservation gives absolute priority to wildlife or one or two aspects only. That simply is not true. We think and worry about the human aspect as much as any other part of society.

W. J. Cairns (hon. conference director). Max Nicholson's theme of the common and indivisible sea between Norway and the UK has been supported by this conference. The common environmental protection case is now launched.

We must now seriously explore alternative instruments for direct and serious environmental collaboration at the extra-governmental level. The Scottish and Norwegian organisers undertake to commit themselves to furthering this responsibility.

[6]Darling, F. F. (Ed.) (1955). *West highland survey: an essay in human ecology.* University of Oxford Press, London, UK. (A study originally commissioned by the Development Commission and the Department of Agriculture for Scotland in 1944.)
[7]Nature Conservancy Council (1974). *Isle of Rhum National Nature Reserve: reserve handbook.* Nature Conservancy Council, North West (Scotland) Region, Inverness, UK.

Glossary of Terms and Abbreviations

Only terms which occur in these proceedings are included here. Wider coverage is given in the Institute of Petroleum publication *Glossary of Petroleum Terms*, the Bank of Scotland Information Service's *The Oil and Gas Industry A Glossary of Terms (1978)* and elsewhere.

°*API*: specific gravity, expressed in degrees of the American Petroleum Institute. °API = 141·5 ÷ SG(60° F/60° F) × 131·5.

blow-out: an (accidental) escape of oil or gas during drilling.

crude oil: see *petroleum*.

downstream: petrochemicals production and petroleum products manufacture; the term may be used comparatively (cf. upstream).

dry gas: see *wet gas*.

feedstock: material to be processed.

gasoline: petroleum distillate used as fuel in spark ignition internal combustion engines.

LNG: liquefied natural gas, gaseous (mainly methane) at normal temperature and pressure, liquefied by high pressure and low temperature to facilitate handling.

LPG: liquefied petroleum gas, gaseous (mainly mixtures of butane and propane) at normal temperature and pressure, liquefied by high pressure or low temperature to facilitate handling.

natural gas: gas occurring under pressure in natural rock formations often with *petroleum* (hence associated gas), classified as *wet* or *dry* according to the proportion of gasoline constituents.

NGL: natural gas liquids: liquids derived from natural gas, including *LPG* and natural gasoline. The term usually excludes *LNG*.

on stream: status of industrial plant in production (converse of 'off stream').

petroleum: crude oil: a naturally occurring liquid mixture of many different hydrocarbons and having a wide range of colors and odors; modern technical use of the term includes gaseous and solid hydrocarbons.

ppb: parts per billion: mg kg^{-1} or mg litre^{-1}.

ppm: parts per million: mg g^{-1} or mg litre^{-1}.

SNG: substitute natural gas.

spud in: to start drilling.

swh: significant wave height: mean height of waves in the highest third of a long sequence of waves.

upstream: exploration and production, but not including refining and subsequent activities; the term may be used comparatively (cf. *downstream*).

wet gas: natural gas may be classified as 'wet' or 'dry' according to whether the proportion of gasoline constituents in it is large or small.

Index

Wildlife
  conservation, 306–7
    see also Conservation
  disturbance, 53
  resources, 44–8
Wilhelmshaven, 257, 258, 259, 287, 288
Working hours, 81

World Conservation Strategy, 50, 54, 307

Ythan Estuary, 92

Zetland County Council Act 1974, 228, 272
Zostera spp., 47